Current Topics in Microbiology and Immunology

Volume 402

More information about this series at http://www.springer.com/series/82

Richard H. ffrench-Constant
Editor

The Molecular Biology of *Photorhabdus* Bacteria

Responsible Series Editor: Klaus Aktories

Springer

Editor
Richard H. ffrench-Constant
Centre for Ecology and Conservation
University of Exeter
Penryn
UK

ISSN 0070-217X ISSN 2196-9965 (electronic)
Current Topics in Microbiology and Immunology
ISBN 978-3-319-52714-7 ISBN 978-3-319-52715-4 (eBook)
DOI 10.1007/978-3-319-52715-4

Library of Congress Control Number: 2017937925

Printed on acid-free paper

This Springer imprint is published by Springer Nature
The registered company is Springer International Publishing AG
The registered company address is: Gewerbestrasse 11, 6330 Cham, Switzerland

Preface

Photorhabdus as a Bug for All Seasons

This book focuses on recent advances in the molecular biology of *Photorhabdus* bacteria and emphasises the application of this new research to agriculture and medicine. *Photorhabdus* has proven to be a 'gold mine' for the discovery of novel toxins and drugs and the volume is divided into three parts which in turn focus on anti-insect virulence and insecticidal toxins, secondary metabolism and the production of novel drugs and the insect immune response to the bacterium. It also includes recent work looking at *P. asymbiotica* which seems to have evolved the ability to infect human hosts as well as insects. We will now look at each section in a little more detail in order to give the reader a 'flavour' of the chapters to come.

Bugs that Kill Bugs

Aside from the use of the entomopathogenic nematode hosts as biocontrol agents the bacteria they vector also represent a novel source of agricultural control agents. Much of the original interest in *Photorhabdus* bacteria therefore came from the search for novel insecticidal toxins The first three chapters look at some of these novel toxins, how they can be discovered and potential mechanisms for their genetic regulation. Sheets and Aktories describe in exquisite detail the discovery, cloning and structure of the novel Toxin complexes or Tc's. Their chapter takes us all the way from the original cloning of the genes that encode these toxins to an atomic resolution of their structure and a description of what is effectively a totally novel secretion system. Dowling then expands on the theme of toxin discovery by describing a range of elegant cell-based screens that can be used to screen genomic libraries of bacteria for novel toxins and drugs. Finally, Givaudan and Lanois finish this section with an overview of how flagellar regulation is deeply linked to the expression of virulence factors (including toxins) in both *Photorhabdus* and *Xenorhabdus*.

Bugs that do Drugs

The second section of the book switches from primary metabolism to secondary metabolism. Here three more chapters survey the array of drugs produced by secondary metabolism, its genetic regulation and the role of small molecules in quorum sensing. Bozhuyuk and coauthors first comprehensively review all of the known small molecules and peptides made by *Photorhabdus* and other entomopathogenic bacteria. These novel drugs are made by large blocks of polyketide synthase (PKS) and non-ribosomal peptide synthase (NRPS) encoding genes that make a startling diversity of new bioactive natural products. Although the role of most of these drugs in the natural history of *Photorhabdus* remains unclear many are already looking promising for uses in agriculture and medicine. The second chapter by Clarke then looks at the complex genetic regulation of these natural products and how an understanding of regulation may gives us clues as to the role of these small molecules in the complex lifecycle of this fascinating bacterium. Finally, Brameyer and Heerman look at two novel quorum sensing mechanisms that affect clumping in the bacteria and therefore may also affect pathogenicity. They describe an array of novel 'LuxR solos' which probably represent the mechanism whereby the bacterium senses the presence of its different hosts (insect, nematode or man).

Bugs that Infect Both Insects and Man

The final sections look at the response of the insect immune system to both the nematode and the bacterium and at the fascinating cases of human infection by *P. asymbiotica*. Eleftherianos and coauthors describe in great detail the immune response that is mounted to the invasion of either the bacteria alone, the nematode alone or both the bacteria and nematode (as would occur in a natural infection). They use state of the art molecular techniques to look at the insect immune response and how the bacterium can escape it. Finally, Gerrard and Stevens give an overview and update of the clinical infections of *P. asymbiotica* reported from both the USA and Australia and Hapeshi and Waterfield then look at the particular genetic mechanisms unique to *P. asymbiotica* that may have allowed for human infection.

Bugs that Just Keep Giving

As a synthesis this book shows that *Photorhabdus* probably has many more gifts to give. For example, most of the screens for novel insecticidal toxins have been directed against lepidopteran caterpillars. So what further novel toxins would we discover if we also screened beetle larvae? How many new drugs remain undiscovered and how do the different bacteria (*Photorhabdus* and *Xenorhabdus*) make them? Are any of these new drugs useful as novel antimicrobials? The list of remaining research questions for these fascinating bacteria are endless and I hope that the chapters that follow will prompt you to ask and investigate many more.

Penryn, UK Richard H. ffrench-Constant

Contents

Part I
Insecticidal Toxins and Virulence Regulation

Insecticidal Toxin Complexes from *Photorhabdus luminescens*

Joel Sheets and Klaus Aktories

Abstract Various bacterial toxins have potent insecticidal activity. Recently, the Toxin complexes (Tc's) of *Photorhabdus* and *Xenorhabdus* species have become an increased focus of current research. These large tripartite toxins with molecular masses >1.4 megadaltons consist of three components termed A, B, and C (or TcA, TcB, and TcC). While TcA is involved in receptor binding and toxin translocation, TcC possesses the specific toxin enzyme activity and TcB is a linker between components TcA and TcC. Here, a structure function analysis of the toxins is described and the application of Tc toxins as potential insecticides is discussed.

Contents

J. Sheets · K. Aktories (✉)
Institute for Experimental and Clinical Pharmacology and Toxicology, Albert-Ludwigs University of Freiburg, Albertstr. 25, 79104 Freiburg, Germany
e-mail: klaus.aktories@pharmakol.uni-freiburg.de

Current Topics in Microbiology and Immunology (2017) 402:3–24
DOI 10.1007/82_2016_55

Published Online: 23 February 2017

1 Introduction

1.1 The Life Cycle of Insecticidal Nematodes

Bacteria from the genus *Photorhabdus* and *Xenorhabdus* are found in symbiotic association with entopathogenic nematodes of the family *Heterorhabditidae* and *Steinernematidae,* respectively. These nematodes are used with limited success as effective biological control agents against important agricultural insect pests of the order Coleoptera and Lepidoptera, including white grubs, weevils, and codling moths (Lacey and Chauvin 1999; Lacey and Georgis 2012). Entopathogenic nematodes employ these bacteria that live within the gut of the nematode to kill their hosts after invading susceptible larvae either through oral, anal, or tracheal pathways. Once the nematodes enter their host, they release their bacteria, which kill the insect usually within 48 h. The rapid kill, along with the production of other virulence factors, helps to defeat the insect's innate immunological defenses allowing the nematodes to survive and reproduce within the insect carcass (Daborn et al. 2001; Silva et al. 2002; Forst and Nealson 1996). The larval carcass provides sufficient nutrients for the nematodes to complete a number of life cycles before the infected carcass ruptures and releases thousands of new nematode progeny that then search for new insect hosts to start the cycle over again (Bowen 2000; Forst et al. 1997). Live entomopathogenic nematodes as insect biological control agents require a narrow effective range of soil moisture, and temperature to work, and demand careful handling and refrigerated storage requirements. These requirements have restricted their effectiveness as biological insect control products and has limited their commercial applications (Lacey and Georgis 2012).

1.2 Specificity of Insecticidal Toxins

Photorhabdus luminescens and *Xenorhabdus nematophilus* bacteria produce a variety of protein toxins, some forming very large (>1.4 MDa) oligomeric tripartite toxin complexes (Tc) that have high levels of toxicity toward insect pests detrimental to important agricultural crops (Waterfield et al. 2001b; ffrench-Constant et al. 2007). Other bacteria, including *Photorhabdus asymbiotica*, *Xenorhabdus bovienii*, *Serratia entomophila*, *Yersinia entomophaga*, *Y. pestis*, and *Pseudomonas syringae pv.* also produce similar toxin complexes (Parkhill et al. 2001; Buell et al. 2003; Wilkinson et al. 2009; Hurst et al. 2000). All of these bacteria represent sources for new genes encoding potent insect toxins that can potentially be used in biological control strategies or in genetically modified plants as alternatives or supplements to insecticidal crystal (Cry) toxins from *Bacillus thuringensis* currently employed in agriculture. Like many Cry proteins, the toxin complex proteins from *Photorhabdus luminescens*, *Xenorhabdus nematophilus*, and *Yersinia entomophaga* appear to be toxic only to certain orders of insects. Although their complete spectra of insecticidal activity have not been fully characterized, the toxin complexes from *Photorhabdus luminescens* are generally very potent against Coleopteran insects such as the Southern corn rootworm (*Diabrotica undecimpunctata howardi*), and the Colorado potato beetle (*Leptinotarsa decemlineata*), which are insect pests of significant agricultural importance (Waterfield et al. 2001a; Bowen and Ensign 1998). The toxin complex proteins from *Xenorhabdus nematophilus* have limited potency against Coleopteran insects, but have potent activity against crop damaging Lepidopteran insects, such as tobacco budworm (*Heliothis virescens*), corn ear-worm, (*Helicoverpa zea*), and beet armyworm (*Spodoptera exigua*) larvae (Morgan et al. 2001; Sergeant et al. 2006). The toxin complexes of *Yersinia entomophaga* have been reported to have oral activity against both Coleoptera and Lepidopteran insects (Hurst J. Bact. 2011). The biochemical source of the specificity is most likely due to specific receptor interactions with the toxin which is known to bind putative receptors located in the membrane of the midgut of susceptible insects (Lee et al. 2007; Sheets et al. 2011). Identification and characterization of these receptors and binding interactions have not been fully reported. Other differences in the biological character of the insect midgut, such as pH, lipids, and presence of different proteases in insects of different orders may also have a role in the specificity of the toxin complexes and for their safety toward other organisms. In addition, all of the toxin complexes so far characterized are active by both oral ingestion and injection into the haemocel of the insect, which implies that the toxin complexes may have target sites that are not simply located in the lumen of the insect midgut (Bowen and Ensign 1998; Bowen 2000).

We will focus on the toxin complexes found in *Photorhabdus luminescens* for the purpose of this chapter, since significant structural and functional information has recently been determined for a subset of Tc's from this bacterium. Where similar data are available for Tc's from other bacteria (esp. *Xenorhabdus nematophilus* and *Yersinia entomophaga*) corresponding comparisons will be made. Understanding how these insect toxins function can assist with the development of new genes for biological or transgenic insect pest control. In addition, the biochemical mechanism of these toxins can provide insight into the mechanism of action of other pathogenic bacteria that cause animal diseases, and provide insight for designing new therapeutic protein reagents.

2 'Toxin Complex' (Tc) Toxins

2.1 *Nomenclature of Tc Toxins*

The nomenclature applied to many of the different Tc genes and subunits is confusing. Initial nomenclature was based upon the order that the genes were found on four (*a*, *b*, *c*, and *d*) different loci in *Photorhabdus luminescens* strain W-14 (Bowen et al. 1998). Hence they were named, for example, *tcaA* for the toxin complex (*tc*) gene on the '*a*' locus encoding the first or '*A*' protein. Later it was determined that all toxin complexes so far examined in detail consist of three functionally different types of proteins or subunits (named A, B, and C), and it became apparent that a unified nomenclature was required (Waterfield 2006). Similar nomenclature was applied for the toxins found in *Xenorhabdus luminescens*, using Xpt standing for *Xenorhabdus* particulate toxin (Sergeant et al. 2003). The *Xenorhabdus* genes were then termed A, B and C based on the order in which they were found in the genome, giving *XptA*, *XptB* and *XptC*. Unfortunately, for both *Photorhabdus* and *Xenorhabdus* Tc proteins, it was then not clear via this nomenclature if we were discussing a type A, B or C protein as defined by its function. For example, tcaC1 is in fact a functional type B protein, tccB a functional type A protein, xptB1 a functional type C protein, and xptC1 is a functional type B protein (Fig. 55.1). The nomenclature for the toxin complex encoding genes isolated from *Yersinia entomophaga* and *Serratia entomophila* was more straight forward, with *yenA and sepA* encoding type A proteins, *yenB* and *sepB*, encoding type B proteins and *yenC* and *sepC* genes encoding type C proteins (Hurst et al. 2000, 2011). Whilst it is not the purpose of this chapter to define a more understandable nomenclature for the toxin complex proteins and genes, it is clear that understanding if the gene encodes a functional type A, B, or C protein is paramount. Knowing what organism the toxin is derived from, and defining a tripartite toxin complex in terms of its bacterial origin and protein composition is thus also important. In this chapter, therefore, we will indicate if a protein is a functional type A (tcA-like), B (tcB-like), or C (tcC-like) protein, along with the bacterial source of the protein and give a unique name to a toxin complex of specifically associated proteins.

2.2 TcA-like Proteins of the Toxin Complex

2.2.1 Interaction of TcA-like Proteins with the Host Cell Membrane

Early biochemical studies on purified toxin complexes isolated from *Photorhabdus luminescens* and *Xenorhabdus nematophilus*, showed toxin complexes to be very large and composed of multiple classes of proteins (Bowen et al. 1998; Bowen and Ensign 1998; Sergeant et al. 2003, 2006). The first measured binding interaction of a TcA-like toxin component with host cells was reported with XptA1. This study showed that the XptA1 protein binds to brush border membrane vesicles (BBMV) prepared from *Pieris brassicae* and to insect derived Sf21 cells, both of which are susceptible to the toxicity of this protein. In contrast, the related XptA2 protein, which is not toxic to these organisms, does not bind to BBMV's from *P. brassicae* or to Sf21 insect cells (Lee et al. 2007). Additional binding interactions were demonstrated for XptA2 using surface plasmon resonance showing that it binds with a K_d of 0.2 nM to solubilized insect midgut BBMV's prepared from *Heliothis zea*, an insect that XptA2 is toxic against (Sheets et al. 2011). These results support earlier studies that the determinate of insect specificity resides with the type of A

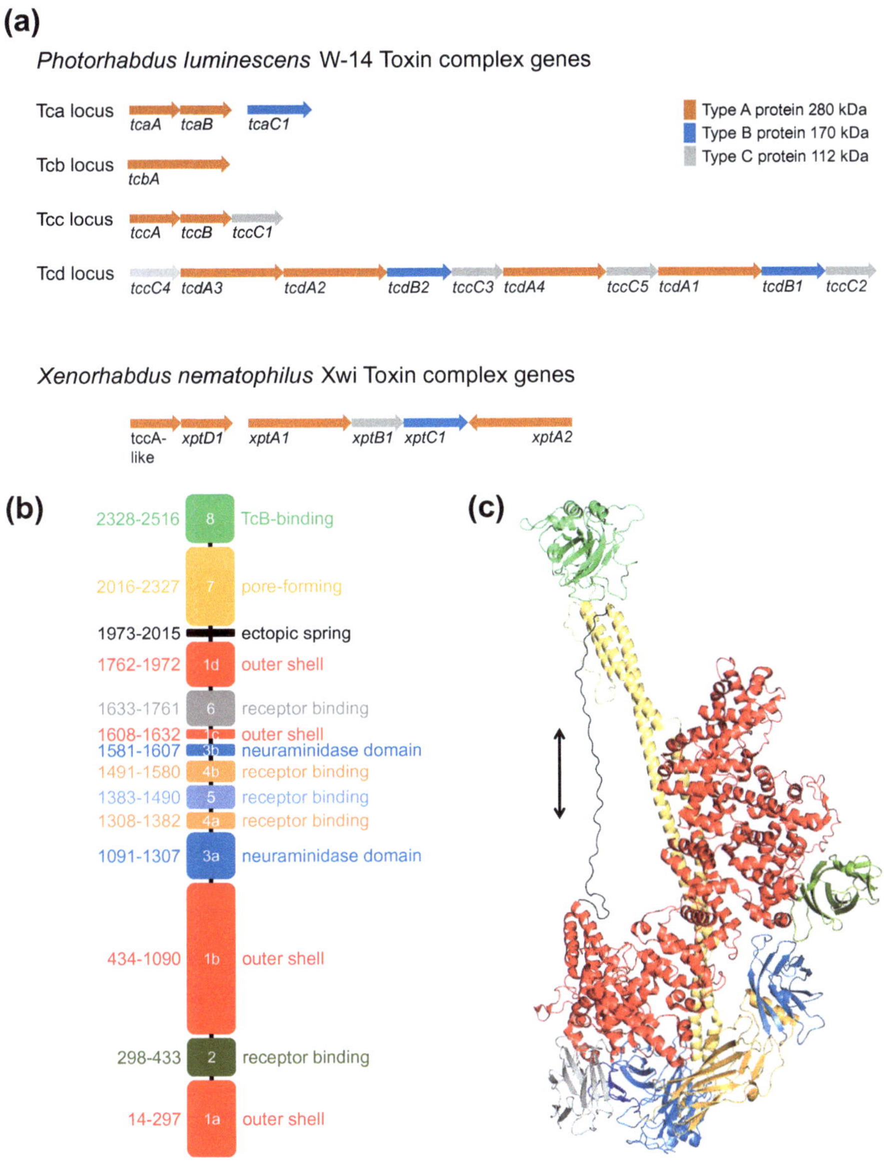

Fig. 55.1 Gene structures of Tc toxins and structure of the TcA-like toxin component TcdA1. (*A*). Gene structure for toxin complexes from *Photorhabdus luminescens* W-14, and *Xenorhabdus nematophilus* Xwi. Genes are colored according to the type of protein they encode (Type A, B, or C-like proteins). Note that tcaA + tcaB; tccA + tccB; and tccA-like + xptD1 are bifurcated type A proteins and presumably are expressed together to form a complete A protein. (*B*). Domain structure of the TcA-like toxin component TcdA1. (*C*). Crystal structure of TcdA1 [PDB 4o9y by PyMOL, modified from Meusch et al. 2014]. The double arrow shows the entropic spring module

protein contained by the toxin complex and that the additional B and C proteins function to potentiate the activity of the A protein (Sergeant et al. 2003; Waterfield et al. 2005).

Additional membrane interactions of the TcA-like protein from *Photorhabdus luminescens* demonstrated the ability of TcdA1 to form pores using $^{86}Rb^{+}$ release experiments on HT-29 cells (Lang et al. 2010). Similarly, TcdA1 forms ion-permeable channels in artificial lipid bilayer membranes with a single channel conductance of 125 pS in 150 mM KCl. The channels formed by TcdA1 were cation selective and the addition of B and C proteins (TcdB2-TccC3) to TcdA1 imbedded in the artificial membrane blocked the conductance of the channel by up to 75%. These results suggest that the binding site for the B-C proteins to the A protein is near the channel formed by the oligomeric A protein (Lang et al. 2013). Similar results were obtained with using the A protein XptA2 from *Xenorhabdus*, where adding XptA2 to one side of the black lipid membrane in the presence of 0.15 M KCl, increased the measured membrane current in a stepwise manner, indicative of formation of single channels with conductance of ~100 pS (Sheets et al. 2011). Meanwhile the interaction of the TcA-like protein TcdA1 with nanodiscs as a membrane model has been reported in near atomic detail (Gatsogiannis et al. 2016) (see below).

2.2.2 The Molecular and Atomic Structure of TcA-like Proteins

TcA-like proteins are large (~280 kDa) and associate into still larger sized oligomers (pentamers) (Figs. 55.1 and 55.2). In some cases, the gene encoding for a TcA protein appears to be bifurcated, presumably encoding the N-terminus and C-terminus portions of a full-sized class A protein by two separate open reading frames (Fig. 55.1a). The structure and organization of bifurcated TcA protein in *Photorhabdus* have not been fully characterized, but it is assumed that the expression of both the genes is required for them to associate into oligomers to produce a fully functional TcA protein. Examples of bifurcated TcA proteins are also known to exist in *Xenorhabdus nematophilus, Yersinia entomophaga,* and other bacteria-containing Tc genes (Morgan et al. 2001; Waterfield et al. 2001b; Hurst et al. 2011).

Recently, structures of a TcA protein from *Photorhabdus luminescens* (TcdA1) in both the pre-pore and pore state have been determined by cryoelectron microscopy (Gatsogiannis et al. 2016; Meusch et al. 2014; Gatsogiannis et al. 2013). TcdA1 forms a 1.41 MDa pentameric bell-shaped structure in the pre-pore state (Fig. 55.2). This structure is similar to the pentameric structure formed by the type A protein (YenA) from *Yersinia entomophaga* (Landsberg et al. 2011) The crystal structure of TcdA1 from *Photorhabdus luminescens* contains 8 different domains in each of the monomers forming the complete pentameric structure (Fig. 55.2). TcA toxin components consists of an inner pore-forming structure, which is built from domains 7 (amino acids 2016–2327) and 8 (amino acids 2328–2516) located at the C-terminus of TcA (Fig. 55.1b, c). Domain 7 consists of two ~200 Å long helices,

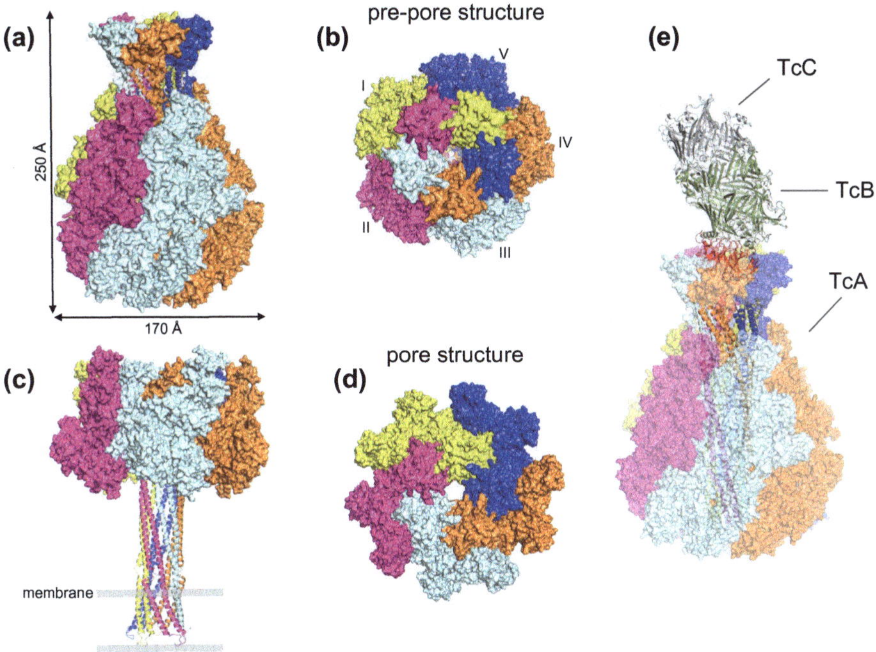

Fig. 55.2 Structures of Tc toxin oligomers as a pre-pore and pore complex. Molecular structure of TcdA1 [modified from Meusch et al. 2014 PDB 4o9y by PyMOL]. (*A*). Side view of pentameric TcdA1 in the pre-pore state. The different colors indicate each protomer. (*B*). Top view of TcdA1 pentamer in the pre-pore state, showing the interaction side with the TcB component. The various protomers are indicated by numbers and color. (*C*). Side view of the pentameric TcdA1 in the pore state, showing the interaction side with the TcB component. (*D*). Top view of the of the pentameric TcdA1 in the pore state. (*E*). Molecular structure of TcdB2-TccC3 on top of the pentameric TcdA1 complex [modified from Meusch et al. 2014]. TcdB2 and TccC3 form a large cage. The N-terminal region of TcdB2 has a β-propeller-like structure and interacts with the platform, which is formed by domain 8 of the TcdA1 pentamer. It is suggested that the ADP-ribosyltransferase domain, which is located in the C_{hvr} region of TccC3, is cleaved from the rest of the protein and lies in the cage in an unfolded conformation. Interaction with the TcA pentamer may open a gate for translocation of the unfolded ADP-ribosyltransferase into the TcA injection machine. TcdA1 is shown as above in a transparent view. (Pictures are modified from Meusch et al. (2014), PDB 409y and PDB4o9x by PyMOL)

one is descending and one is ascending) and both are connected by a loop region (residues 2140–2155). Together with the other eight helices from the other four protomers form the channel of the pore. The pore possesses a funnel-like region at the top, which is formed by domain 8 oligomers (Figs. 55.1 b, c and 55.2). This region builds the docking platform for the interaction with component TcB of the holotoxin (Fig. 55.2 b, d). The channel of TcA is encased by an outer shell, consisting of an all-helical domain (Fig. 55.1 b, c, regions 1a–1d), which has insertions for four putative receptor binding domains (domains 2, 4, 5 and 6 in Fig. 55.1 b, c) and one neuroaminidase-like domain (domain 3 in Fig. 55.1 b, c).

Like many other pore-forming toxins, the structure of TcdA1 can be described as a pre-pore and as a pore state. The pore state of TcdA1 (Fig. 55.2c, d) was obtained by reconstituting the complex into liposomes and nanodiscs, determining its structure by Cryo-EM and single-particle analysis. Transitioning from the pre-pore to pore state results in a large 12 nm movement of residues forming the central channel that is present in the interior of the toxin complex shielded by the outer surface of the toxin (Fig. 55.2). This corresponds to insertion of the α-helical pore into the membrane of the insect midgut, establishing a syringe-like mechanism that could be used to inject proteins into cells. The energy for the insertion event is entropic and is achieved through the action of a linker composed of 48 amino acid residues that connect the shell domains with the central channel-forming part of TcA (Fig. 55.1 b, c). This linker comprises an 'entropic spring' by changing from an extended to a compacted conformation thereby driving the central channel into the membrane. Based upon this structure, the TcA protein primarily functions as a syringe-like delivery system for the toxin complex (Meusch et al. 2014; Gatsogiannis et al. 2013, 2016). The interior lumen of the central channel is initially closed in the pre-pore state, but opens when in the pore state, allowing for injection of proteins through the channel into the insect gut (Gatsogiannis et al. 2016). Although the channel opens when in the pore state, it is still sufficiently narrow that it would require unfolding of the proteins to allow their passage through the pore and into a cell (Gatsogiannis et al. 2013). The structure shows how TcdA1 by itself can function in both a pore-forming role, and also as a protein toxin delivery system.

The identity of the receptors in the insect midgut that bind the toxin are not known, but the structure of the toxin complex in the pore state suggests that the receptor binding domains are about 125 Å from the surface of the membrane, requiring the receptors on the midgut cell surface to be both large and elongated to sufficiently interact with the receptor domains on the toxin complex (Meusch et al. 2014).

2.2.3 Is TcA per se Insecticidal?

Early insect bioassay studies suggested that the TcA protein (TcdA1) was the 'toxin' portion of the Tc, and that the addition of co-expressed TcB and TcC proteins function to 'potentiate' the activity of the TcA protein (Waterfield et al. 2005; Sergeant et al. 2003). Indeed, when the gene encoding TcdA1 from *Photorhabdus luminescens* was inserted into the *Arabidopsis thaliana* genome the resulting transgenic plants were toxic to tobacco hornworms (*Manduca sexta*), and inhibited the growth of southern corn rootworms (*Diabrotica undecimpunctata howardi*) (Liu et al. 2003). The ability of TcdA1 to bind and form pores in insect midgut membranes most likely accounted for the insecticidal activity of the TcdA1 expressed by the transgenic plants. Later studies showed that both the TcB and TcC proteins are needed for the toxin complex to exhibit its full insecticidal activity (Waterfield et al. 2005).

2.3 TcB-like Proteins of the Toxin Complex

TcB-like proteins are about 170 kDa in size and bind to both the TcA and TcC proteins to facilitate the formation of a complete Tc toxin. TcB proteins have not shown any toxicity against insects and little is known about their function as part of the toxin complex other than to connect the TcC protein to the TcA complex. They can be singly expressed as soluble proteins, but in practice are often co-expressed with the TcC protein or expressed as a fusion protein with the TcC protein attached to its C-terminus. The co-expression of both TcB and TcC proteins together facilitates the successful expression of TcC proteins, which are often difficult to express individually and typically form an insoluble full-length protein that poorly associates with the TcB protein. For this reason, it has been thought that the TcB protein may act as a chaperon for the TcC protein. Interestingly, genes encoding naturally fused TcB–TcC proteins have been found in some bacteria with toxin complex genes (Yang and Waterfield 2013). The crystal structure of two TcB–TcC fusion proteins (TcdB2–TccC3) from *Photorhabdus luminescens* has been solved at 2.35 Å resolution and shows TcdB2 having a large hollow cocoon-like structure (Meusch et al. 2014). The structure is highly similar to the structure of the TcB–TcC components of *Yersinia entomophaga* (Busby et al. 2013). The N-terminus portion of the TcB protein binds to the TcA protein and forms a β-propeller type structure that is positioned on top of the channel formed by the TcA protein (Meusch et al. 2014; Busby et al. 2013). The β-propeller is in an open conformation when the TcB–TcC protein is bound to the TcA protein allowing for a clear channel through the toxin complex (Meusch et al. 2014).

2.4 TcC-like Proteins of the Toxin Complex

The TcC-like proteins are about 112 kDa in size and are important for potent toxicity of the complex. As mentioned above, TcC proteins need to be co-expressed with the TcB protein in *E. coli* or other protein expression systems to properly associate with the TcA protein for production of a complete and biologically active toxin complex (Sergeant et al. 2003). For the toxin complexes from *Photorhabdus* and *Xenorhabdus* studied in detail so far, when the B and C proteins are co-expressed, or expressed as a B-C fusion product, the two proteins are isolated as a soluble binary complex where the C protein is cleaved into two parts as part of an apparent activation mechanism (Lang et al. 2010; Sheets et al. 2011). The structure of the C protein can be considered to be composed of two domains. The core domain comprises the predominant N-terminal section of the protein and contains *rearrangement hotspot* (RHS) repeats (Busby et al. 2013). These repeats are believed to form an aspartyl autoprotease in the C2 protein of *Y. entomophaga* and in TccC3 of *P. luminescens* (Meusch et al. 2014; Busby et al. 2013). Thus, the TcC protein autoproteolyses when co-expressed with the TcB protein to form two separate proteins. The site of cleavage occurs approximately two-thirds down the protein toward the C-terminus at a highly conserved junction among different TcC-like proteins. Immediately after the site of cleavage, the amino acid sequences of different TcC proteins from various organisms radically diverge, and this domain referred to as the hypervariable region (hvr). Cleavage of the hvr from the TcC protein releases a protein fragment (TcC_{hvr}) of about 32 kDa that non-covalently associates with the TcB–TcC protein. As describe below, TcC_{hvr} contains enzymatic activity that defines the toxic activity of the toxin complex.

2.5 Structure of the Complete TC Complex

A complete toxin complex that is fully active against insects is composed of all three Tc proteins (TcA, TcB, and TcC) where the TcA protein oligomerizes into a pentamer in most cases, and binds a TcB protein which in turn is bound to a TcC protein. Although the C protein is cleaved in the complete toxin complex, the cleaved TcC_{hvr} must remain associated with the toxin complex for it to be active against insects. Mutant TcB–TcC fusion proteins having a deleted TcC_{hvr} still bind tightly to the TcA protein and form a complex, but the toxin complex lacks potent insecticidal activity (Busby et al. 2013). The crystal structure of TcdB2–TccC3 presents a hollow cocoon structure (Fig. 55.2e), similar to that described for Tc YenB–YenC2 proteins described at 2.35 Å resolution (Busby et al. 2013). The TcC_{hvr} protein is assumed to be encapsulated within the cocoon structure of the TcdB2–TccC3 complex in a non-covalent manner possibly unfolded due to the probable hostile environment composed of positive charges and hydrophobic patches in the interior of the B-C structure (Meusch et al. 2014).

Recently, an elegant body of X-ray crystallography and electron cryomicroscopy structural studies at 4 Å resolution has been reported for crystals of TcdA1 and of a complete *Photorhabdus* toxin complex (PTC3), having TccC3 as the type of C-protein in the complex. The structure was obtained through co-crystallization of purified A protein TcdA1 and the binary B-C proteins TcdB2-TccC3 expressed as a fused single gene product (Meusch et al. 2014). The binary TcdB2–TccC3 proteins bind tightly to the top of the TcdA1 pentamer at an approximate 45° angle to its longitudinal axis forming a distorted six-bladed β-propeller structure having pseudo fivefold symmetry (Fig. 55.2e). The β-propeller is in an open conformation when bound to the TcdA1 pentamer, providing a chamber for the C_{hvr} to pass through into the channel of the pentameric TcdA1, as evidenced by extra density found inside the pore of the toxin complex (Meusch et al. 2014). It is believed that C_{hvr} is in an unfolded state when inside the channel. The structure of the entire PTC3 complex has an analogous overall structure to the complete pentameric toxin complex from *Yersinia entomophaga*, except that the *Yersinia* toxin complex (Yen-Tc) is reported to be associated with two different chitinase proteins (Chi1 and Chi2) bound to the outer surface of each YenA protein for a total of ten chitinases, along with two type C proteins (C1 and C2) associated with a single B protein (Landsberg et al. 2011). In addition, the A proteins of YenTc are encoded by two separate genes that essentially form a single TcdA-like protein. Although early structural studies indicated two C proteins were present for each B protein, later crystal structures of the BC proteins of YenTc showed them to form binary complexes very similar as seen for TccB2–TccC3 of *Photorhabdus* (Busby et al. 2013). Thus we suspect that the purified toxin complex protein reported (Landsberg et al. 2011) was most likely composed of two separate TCs, each sharing the same A and B proteins, but having different C proteins. Such a structure would conform to the A_5BC structure observed for PTC3 (Meusch et al. 2014).

The role of the chitinases in YenTc was originally thought to function to degrade the chitin-rich peritrophic membrane of the insect gut to facilitate entry of the toxin into the cells. Such a mechanism is not required of Tc's from *Photorhabdus* or *Xenorhabdus* since these Tc's do not contain chitinases yet are highly toxic toward insects. Chitinase genes have been found in close proximity with the toxin complex genes of *Xenorhabdus nematophilus*, but are not required for their biological activity (Morgan et al. 2001). Later characterization of the chitinases of YenTc showed optimum endochitinase activity at pH values of neutral to acidic, whereas the pH of the lumen of the gut of some insects that the toxin was active against was as high as 9, where these chitinases were found not to have any biological activity (Busby et al. 2012). The chitinases found on YenTc may instead have a role in toxicity after the complexes enter the cell and kills the insect. Certainly, making and characterizing YenTc mutants lacking the chitinase proteins would bring clarity to the role of these proteins in the toxicity mechanism.

The genomes of *Photorhabdus luminescens* contains multiple different genes encoding TcA–, TcB–, and TcC–like proteins, which allows for the formation of different toxin complexes depending upon which A, B, and C protein combinations are contained within the complex (Waterfield et al. 2005). Different TcB and TcC proteins can be combined with different TcA proteins to form a wide variety of different Tc's which can express different toxicity mechanisms and attack different hosts. Hybrid toxins can potentially be made that have unique insect specificity depending on the type of TcA, TcB, and TcC proteins contained in the toxin complex. Although a complete comparison has not been made, it has been shown that different pairs of TcB and TcC proteins can be co-expressed and they bind to the various TcA-like proteins to form mixtures of different toxin complexes. Indeed, TcB and TcC proteins from *Photorhabdus luminescens* (TccB2 and TccC3) have been co-expressed and added to a TcA protein from *Xenorhabdus nematophilus* (XptA2) to form a hybrid toxin complex from two different organisms. This toxin complex expresses the insect selectivity determined by XptA2, but the molecular toxicity is a result of the actions of the hypervariable region of TccC3 (Sheets et al. 2011).

3 Biological Activities of Tc Toxins

Early studies indicated that the maximal insect-killing activity of Tc toxins depends on the combination of all three Tc components, including TcA, TcB, and TcC (Waterfield et al. 2001a). The tripartite *Photorhabdus luminescens* Tc complexes were studied in hemocytes obtained from the hemocoel of *Galleria mellonella* larvae (Lang et al. 2010). These studies revealed inhibition of phagocytosis by the toxin complexes consisting of TcdA1, TcdB2, and TccC3 (PTC3) or TcdA1, TcdB2 and TccC5 (PTC5), while the individual proteins were without effects. It turned out that inhibition of phagocytosis was caused by effects of the toxin on the actin cytoskeleton. However, studies with hemocytes and also with mammalian cells showed that PTC3 and PTC5 each affect the actin cytoskeleton in a different manner. While PTC3 induces aggregation of F actin, PTC5 causes formation of stress fibers. The combination of both toxins results in strong clustering of F-actin accumulations with loss of the normal cell morphology. The underlying molecular mechanisms are the ADP-ribosyltransferase activities of the TccC3 and TccC5 components (Lang et al. 2010). However, the toxin targets and the functional consequences of toxin catalyzed ADP-ribosylation are different for TccC3 and TccC5 toxin components.

3.1 TccC3 Is an ADP-Ribosyltransferase that Targets Actin

In spite of a low overall sequence homology between most bacterial ADP-ribosylating toxins, their molecular structures are very similar (Hottiger et al. 2010; Simon et al. 2014; Fieldhouse and Merrill 2008; Vogelsgesang et al. 2007; Pinto and Schuler 2015). Moreover, essential amino acid residues critical for catalysis of the ADP-ribosyltransferase reaction are frequently conserved. Typical for a large group of bacterial ADP-ribosyltransferases is the so-called "RSE" motif, which represents three amino acids involved in NAD-binding (residues "R" and "S") and in catalysis (catalytic glutamate "E") (Hottiger et al. 2010). TccC3, which harbors the ADP-ribosyltransferase activity in its C_{hvr}, belongs to this family of RSE enzymes (Pfaumann et al. 2015). Mutational analyses revealed that Arg791 and Glu943 are the pivotal residues of the typical RSE motif. The crucial serine residue has not been determined with absolute certainty, however serine866 or serine871 are good candidates (Pfaumann et al. 2015).

3.1.1 Modification of Actin by TccC3 at Threonine148

TccC3 mono-ADP-ribosylates actin at threonine148 (Lang et al. 2010). Modification of actin at this site has no effect on the native state of actin, thus ATP-binding is preserved and modified actin still inhibits DNAse I (Lang et al. 2016).

Moreover, the critical concentration for actin polymerization is not changed. However, threonine148 is located at the site where actin interacts with thymosin-β4 (Lang et al. 2010). The actin-binding protein thymosin-β4 ($\sim$5 kDa) belongs to a group of rather small peptides, consisting of 42–45 residues, which sequester G-actin and inhibits salt-induced actin polymerization (Mannherz and Hannappel 2009). ADP-ribosylation of actin at T148 blocks the interaction of thymosin-β4 with actin and prevents its inhibition of actin polymerization (Lang et al. 2010). Other actin-binding proteins that interact with actin involve threonine148 (Lang et al. 2016). For example, the interaction of gelsolin and of ADF/cofilin with actin ADP-ribosylated at threonine148 is impaired. Under physiological conditions, gelsolin and ADF/cofilin induce fragmentation of F-actin and thereby control the dynamics of actin treadmilling (Pollard and Cooper 2009; Dominguez and Holmes 2011). By contrast, the interaction of actin ADP-ribosylated at threonine148 with profilin is not affected. Profilin supports plus-end polymerization of actin and enhances formin-induced actin elongation. Thus, ADP-ribosylation of actin at threonine148 favors actin polymerization while actin fragmentation and severing is inhibited. Moreover, ADP-ribosylation at threonine148 appears to affect actin–actin interaction in favor of F-actin bundling and aggregation, eventually, resulting in destruction of the normal cell morphology and in cell death (Lang et al. 2010; Lang et al. 2016) (Fig. 55.3b).

3.1.2 *Photorhabdus* Toxin Photox Modifies Actin at Arginine177

Various bacterial toxins modify actin by ADP-ribosylation (Aktories et al. 2011) including the binary toxins *Clostridium botulinum* C2 toxin (Aktories et al. 1986), *C. perfringens* iota toxin (Schering et al. 1988), *C. difficile* transferase CDT (Gülke et al. 2001; Schwan et al. 2009) and *Bacillus cereus* vegetative insecticidal proteins (VIP) (Han et al. 1999). Moreover, various bacterial effectors like *Salmonella enterica* SpvB (Tezcan-Merdol et al. 2001) and *Aeromonas salmonicida* AexT (Vilches et al. 2008), which probably enter target cells by type III secretion, ADP-ribosylate actin. However, these toxins do not modify threonine148 but instead ADP-ribosylate actin at arginine177 (Vandekerckhove et al. 1987, 1988; Hochmann et al. 2006). This modification of actin at arginine177 inhibits the formation of actin filaments by steric hindrance and therefore blocks actin polymerization (Aktories et al. 1986; Aktories and Wegner 1989, 1992). Moreover, it causes F-actin depolymerization by an actin-capping effect at the plus-ends of actin filaments (Wegner and Aktories 1988). *Photorhabdus luminescens* produces a potential toxin called Photox, which possesses ADP-ribosyltransferase activity and modifies actin at arginine-177 (Visschedyk et al. 2010). Although the translocation mechanism of Photox into target cells is not known, it is remarkable that *Photorhabdus luminescens* produces toxins that cause polymerization of actin (e.g., TccC5-containing Tc complexes) and also release toxins (e.g., Photox) that induce depolymerization of actin filaments and of the actin cytoskeleton. Eventually, both toxin effects are detrimental for targeted cells. However, a precise spatial—and

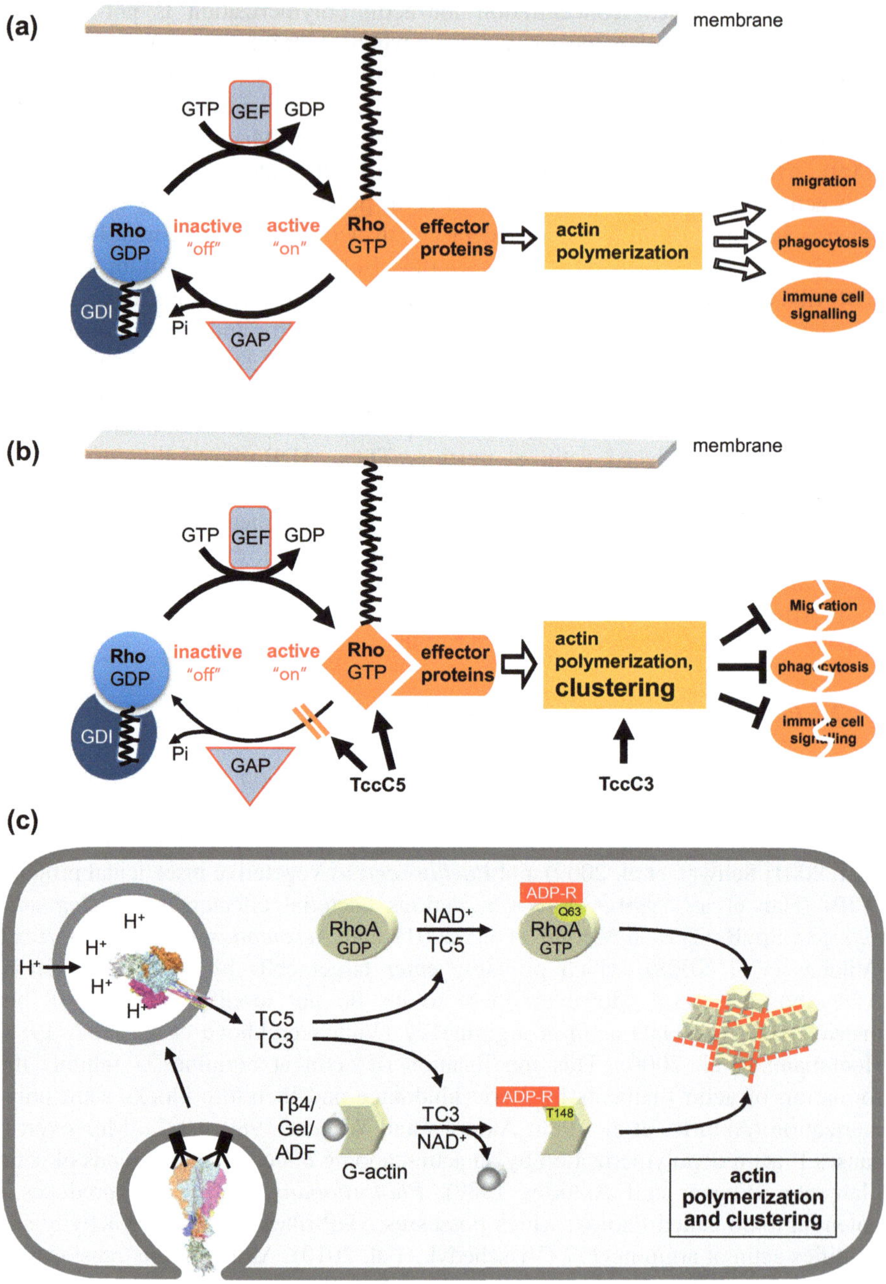
(a)
membrane
GTP
GEF
GDP
Rho
GDP
GDI
inactive
"off"
active
"on"
Pi
GAP
Rho
GTP
effector proteins
actin polymerization
migration
phagocytosis
immune cell signalling
(b)
membrane
GTP
GEF
GDP
Rho
GDP
GDI
inactive
"off"
active
"on"
Pi
GAP
Rho
GTP
effector proteins
actin polymerization, clustering
TccC5
TccC3
(c)
H+
RhoA
GDP
NAD+
TC5
ADP-R
Q63
RhoA
GTP
TC5
TC3
Tβ4/
Gel/
ADF
G-actin
TC3
NAD+
ADP-R
T148
actin polymerization and clustering

◂**Fig. 55.3** Mode of action of Tc toxins. (*A*) TccC5 activates Rho proteins by ADP-ribosylation at glutamine63. **a** Rho proteins are regulated by a GTPase cycle, are activated by GEFs and inactivated by GTP hydrolysis, which is facilitated by GAPs (GTPase activating proteins). GDIs (guanine nucleotide dissociation inhibitors) keep Rho proteins in the cytosol. GTP-bound Rho activates multiple effectors. A major effect of active Rho is the organization of the actin cytoskeleton involved in phagocytosis, migration and immune cell signaling. **b** TccC5 ADP-ribosylates RhoA in glutamine63, thereby GTP hydrolysis is blocked and Rho is persistently active, resulting in enhanced actin polymerization and deregulation of actin functions. In addition, TccC3 ADP-ribosylates actin causing actin clustering (see below). (*B*). Model for the action of *Photorhabdus* Toxin complex. The toxin complex proteins come together to form a complete toxin complex which binds to membrane bound receptors on midgut cells and is endocytosed into the cell. At low pH in the endosome, the Tc forms a pore that releases, through a syringe-like mechanism, the C_{hvr} of the TcC protein into the cytosol. The C_{hvr} of TccC3 ADP-ribosylates actin at threonine-148, which inhibits actin interaction with actin-binding proteins like thymosin β-4 (Tβ-4), gelsolin (Gel) or ADF and induces actin polymerization and clustering. The C_{hvr} of TccC5 ADP-ribosylates Rho GTPases resulting in its persistent activation leading to actin polymerization. Both toxins largely redistribute the actin cytoskeleton and cause actin clustering leading to toxicity

time-dependent control of the toxins' actions may provide an advantage for *Photorhabdus luminescens* and its specific life cycle.

3.2 *TccC5 Is an ADP-Ribosyltransferase that Targets Rho Proteins*

TccC5 also harbors ADP-ribosyltransferase activity in its C_{hvr}. The RSE motif for TccC5 ADP-ribosyltransferase activity comprises arginine774, serine809 and glutamate886. Mutations of these residues result in inhibition of the ADP-ribosyltransferase activity (Pfaumann et al. 2015). Biochemical studies and mass spectrometric analysis showed that Rho proteins are the substrates of TccC5 (Lang et al. 2010).

3.2.1 Rho Proteins Are Substrates of TccC5

Rho proteins belong to the superfamily of Ras proteins and are regulated by a GTPase cycle (Cherfils and Zeghouf 2013; Jaffe and Hall 2005; Heasman and Ridley 2008) (Fig. 55.3a). They are active in the GTP-bound form and inactive after GTP hydrolysis. GTP-binding is achieved by guanine nucleotide exchange factors (GEFs) (Garcia-Mata and Burridge 2007), which induce the release of GDP bound to Rho proteins. Because the concentration of GTP in the cytosol is higher than the concentration of GDP, release of GDP results in subsequent binding of GTP. The activate state of Rho proteins is turned off by hydrolysis of bound GTP, a process which is facilitated by GTPase activating proteins (GAPs) (Scheffzek et al. 1998). In the GTP-bound form, Rho proteins interact with numerous effectors including various protein kinases, adaptor, and regulatory proteins (Thumkeo et al.

2013). A prominent role of Rho proteins is the regulation of the actin cytoskeleton and of actin-dependent processes. Thus, Rho proteins regulate multiple cellular motile functions like cell migration, phagocytosis, vesicle traffic and cytokinesis. However, Rho proteins are also involved in transcriptional control and cell cycle regulation (Hanna and El-Sibai 2013). Moreover, Rho proteins are preferred substrates of bacterial protein toxins and effectors (Aktories 2011; Lemichez and Aktories 2013).

3.2.2 Functional Consequences of Rho ADP-Ribosylation by TccC5

Preferred in vitro substrates of TccC5 are RhoA, RhoB and RhoC, Rac1, 2 and 3, Cdc42 and Tc10 (Pfaumann et al. 2015; Lang et al. 2010). Modification occurs at glutamine63 of RhoA and glutamine61 of Rac and Cdc42. This residue is located in the switch II region of the small GTPases and is essential for GTP hydrolysis (Cherfils and Zeghouf 2013). Thus, toxin-induced ADP-ribosylation at this position inhibits GTP hydrolysis by the small GTPase and turns the GTP-binding protein into a persistently active protein (Lang et al. 2010) (Fig. 55.3a). The consequence of toxin-caused activation of RhoA is the strong formation of stressfibres, which is a typical cell culture response following RhoA activation. Thus, in cell culture, the RhoA effect appears to be dominant. However, when both active TcC components (e.g., TccC3 and TccC5) were added to cell cultures as part of the complete Tc complexes, stressfibers are no longer visible but gross actin clustering is observed, indicating that modification of threonine148 by TccC3 is dominant (Lang et al. 2010). Activation of Rho and Rac proteins by TccC5 was studied in comparison with the cytotoxic necrotizing factor CNF1 of *Escherichia coli*, which activates Rho proteins by deamidation of glutamine63 (Schmidt et al. 1997). Surprisingly, activation of RhoA by deamidation is transient and returns to normal Rho activation levels after 18 h, while TccC5 causes RhoA activation for more than 18 h (Pfaumann et al. 2015). Similar results are observed with Rac proteins. Activation of Rac by CNF1 is transient with subsequent degradation of Rac, while TccC5 causes persistent activation over 18 h without degradation of Rac.

In congruence with TccC3, which directly ADP-ribosylates actin at threonine148, ADP-ribosylation of RhoA by TccC5 favors actin polymerization (Lang et al. 2010; Pfaumann et al. 2015). Accordingly, both toxins share a similar effect on gene transcription regulation by myocardin-related transcription factor A (termed MAL or MRTF). MAL is an essential coactivator of the transcription factor *serum responsive factor* (SRF) (Medjkane et al. 2009; Posern and Treisman 2006). The transcriptional regulator Mal interacts with G-actin thereby inhibiting its transcriptional activity. Polymerization of actin releases MAL from G actin and supports MAL-mediated transcriptional activation resulting in increased expression of multiple proteins involved in organization of the cytoskeleton, adhesion and in motile functions including actin itself (Medjkane et al. 2009; Posern and Treisman 2006). Studies showed that TccC3 as well as TccC5 strongly induce MAL activation (Pfaumann et al. 2015). Thus, manipulation of MAL-dependent function including signaling, migration and adhesion of *Photorhabdus* toxins target cells may play an important role in host–pathogen interaction in larvae.

4 Conclusions

Recent studies have shed light on the structure and function of the huge and highly complex toxin complex toxins from *Photorhabdus luminescens*. These tripartite toxins are a paradigm of a novel toxin delivery system using a syringe-like mechanism to transport an enzymatic toxin into a cell. Because Tc toxins are found in various species from different bacterial genera, these recent findings will help us better understand toxin-dependent host–pathogen interactions caused by these proteins. There are still many outstanding questions concerning the binding of the TcA component of the toxin complex to target cells, the delivery into, and the action of the biological active components (TcC) in host cells. For example, identification of the toxin receptor (or receptors) would be of major importance. Moreover, numerous TcC components have been identified in the genome of *Photorhabdus*, and related bacteria, which share high sequence similarity in their N-terminus but largely differ at the C-terminus, where the biological activity is located. Sequence analyses clearly show that these hypervariable region most likely harbor different biological activities. Therefore, it would be exciting to analyze these activities in detail. Insecticidal bacterial protein toxin play a still increasing role in pest control and parasite management. The various toxins of *Photorhabdus* species including the Tc toxins may increase our armamentarium against various types of insect pests. One prerequisite for the efforts to obtain novel, even more effective insecticidal agents against insect pests is the detailed understanding of the mode of action of these toxins.

References

Aktories K (2011) Bacterial protein toxins that modify host regulatory GTPases. Nat Rev Microbiol 9(7):487-498. doi: 10.1038/nrmicro2592, [pii] nrmicro2592

Aktories K, Bärmann M, Ohishi I, Tsuyama S, Jakobs KH, Habermann E (1986) Botulinum C2 toxin ADP-ribosylates actin. Nature 322:390–392

Aktories K, Lang AE, Schwan C, Mannherz HG (2011) Actin as target for modification by bacterial protein toxins. FEBS J 278(23):4526–4543. doi:10.1111/j.1742-4658.2011.08113.x

Aktories K, Wegner A (1989) ADP-ribosylation of actin by clostridial toxins. J Cell Biol 109:1385–1387

Aktories K, Wegner A (1992) Mechanisms of the cytopathic action of actin-ADP-ribosylating toxins. Mol Microbiol 6:2905–2908

Bowen D, Rocheleau TA, Blackburn M, Andreev O, Golubeva E, Bhartia R (1998) Insecticidal toxins from the bacterium *Photorhabdus luminescens*. Science 280(5372):2129–2132

Bowen DJ, Ensign JC (1998) Purification and characterization of a high-molecular-weight insecticidal protein complex produced by the entomopathogenic bacterium photorhabdus luminescens. Appl Environ Microbiol 64(8):3029–3035

Buell CR, Joardar V, Lindeberg M, Selengut J, Paulsen IT, Gwinn ML, Dodson RJ, Deboy RT, Durkin AS, Kolonay JF, Madupu R, Daugherty S, Brinkac L, Beanan MJ, Haft DH, Nelson WC, Davidsen T, Zafar N, Zhou L, Liu J, Yuan Q, Khouri H, Fedorova N, Tran B, Russell D, Berry K, Utterback T, Van Aken SE, Feldblyum TV, D'Ascenzo M, Deng WL, Ramos AR, Alfano JR, Cartinhour S, Chatterjee AK, Delaney TP, Lazarowitz SG, Martin GB, Schneider DJ, Tang X, Bender CL, White O, Fraser CM, Collmer A (2003) The complete genome sequence of the *Arabidopsis* and tomato pathogen *Pseudomonas* syringae pv. tomato DC3000. Proc Natl Acad Sci USA 100(18):10181–10186. doi:10.1073/pnas.1731982100

Busby JN, Landsberg MJ, Simpson RM, Jones SA, Hankamer B, Hurst MR, Lott JS (2012) Structural analysis of Chi1 Chitinase from Yen-Tc: the multisubunit insecticidal ABC toxin complex of *Yersinia entomophaga*. J Mol Biol 415(2):359–371. doi:10.1016/j.jmb.2011.11.018

Busby JN, Panjikar S, Landsberg MJ, Hurst MR, Lott JS (2013) The BC component of ABC toxins is an RHS-repeat-containing protein encapsulation device. Nature 501(7468):547–550. doi: 10.1038/nature12465, [pii] nature12465

Cherfils J, Zeghouf M (2013) Regulation of small GTPases by GEFs, GAPs, and GDIs. Physiol Rev 93(1):269–309. doi: 10.1152/physrev.00003.2012, [pii] 93/1/269

Daborn PJ, Waterfield N, Blight MA, Ffrench-Constant RH (2001) Measuring virulence factor expression by the pathogenic bacterium *Photorhabdus luminescens* in culture and during insect infection. J Bacteriol 183(20):5834–5839. doi:10.1128/JB.183.20.5834-5839.2001

Dominguez R, Holmes KC (2011) Actin structure and function. Annu Rev Biophys 40:169–186. doi:10.1146/annurev-biophys-042910-155359 [doi]

ffrench-Constant R, Waterfield N (2006) An ABC guide to the bacterial toxin complexes. Adv Appl Microbiol 58:169–183

ffrench-Constant RH, Bowen DJ (2000) Novel insecticidal toxins from nematode-symbiotic bacteria. Cell Mol Life Sci 57(5):828–833

ffrench-Constant RH, Dowling A, Waterfield NR (2007) Insecticidal toxins from *Photorhabdus* bacteria and their potential use in agriculture. Toxicon 49 (4):436–451

Fieldhouse RJ, Merrill AR (2008) Needle in the haystack: structure-based toxin discovery. Trends Biochem Sci 33(11):546–556. doi: 10.1016/j.tibs.2008.08.003, [pii] S0968-0004(08)00188-6

Forst S, Dowds B, Boemare N, Stackebrandt E (1997) *Xenorhabdus* and *Photorhabdus* spp.: bugs that kill bugs. Annu Rev Microbiol 51:47–72

Forst S, Nealson K (1996) Molecular biology of the symbiotic-pathogenic bacteria *Xenorhabdus* spp. and *Photorhabdus* spp. Microbiol Rev 60(1):21–43

Garcia-Mata R, Burridge K (2007) Catching a GEF by its tail. Trends Cell Biol 17(1):36–43. doi: 10.1016/j.tcb.2006.11.004, [pii] S0962-8924(06)00326-6

Gatsogiannis C, Lang AE, Meusch D, Pfaumann V, Hofnagel O, Benz R, Aktories K, Raunser S (2013) A syringe-like injection mechanism in *Photorhabdus luminescens* toxins. Nature 495 (7442):520–523. doi: 10.1038/nature11987, [pii] nature11987

Gatsogiannis C, Merino F, Prumbaum D, Roderer D, Leidreiter F, Meusch D, Raunser S (2016) Membrane insertion of a Tc toxin in near-atomic detail. Nat Struct Mol Biol. doi:10.1038/nsmb.3281

Gülke I, Pfeifer G, Liese J, Fritz M, Hofmann F, Aktories K, Barth H (2001) Characterization of the enzymatic component of the ADP-ribosyltransferase toxin CDTa from *Clostridium difficile*. Infect Immun 69(10):6004–6011

Han S, Craig JA, Putnam CD, Carozzi NB, Tainer JA (1999) Evolution and mechanism from structures of an ADP-ribosylating toxin and NAD complex. Nature Struct Biol 6:932–936

Hanna S, El-Sibai M (2013) Signaling networks of Rho GTPases in cell motility. Cell Signal 25 (10):1955–1961. doi: 10.1016/j.cellsig.2013.04.009, [pii] S0898-6568(13)00124-1

Heasman SJ, Ridley AJ (2008) Mammalian Rho GTPases: new insights into their functions from in vivo studies. Nat Rev Mol Cell Biol 9(9):690–701. doi: 10.1038/nrm2476, [pii] nrm2476

Hochmann H, Pust S, von FG, Aktories K, Barth H (2006) *Salmonella enterica* SpvB ADP-ribosylates actin at position arginine-177-characterization of the catalytic domain within the SpvB protein and a comparison to binary clostridial actin-ADP-ribosylating toxins. Biochemistry 45(4):1271–1277
Hottiger MO, Hassa PO, Luscher B, Schuler H, Koch-Nolte F (2010) Toward a unified nomenclature for mammalian ADP-ribosyltransferases. Trends Biochem Sci 35(4):208–219. doi: 10.1016/j.tibs.2009.12.003, [pii] S0968-0004(09)00242-4
Hurst MR, Glare TR, Jackson TA, Ronson CW (2000) Plasmid-located pathogenicity determinants of *Serratia entomophila*, the causal agent of amber disease of grass grub, show similarity to the insecticidal toxins of *Photorhabdus luminescens*. J Bacteriol 182(18):5127–5138
Hurst MR, Jones SA, Binglin T, Harper LA, Jackson TA, Glare TR (2011) The main virulence determinant of *Yersinia entomophaga* MH96 is a broad-host-range toxin complex active against insects. J Bacteriol 193(8):1966–1980. doi:10.1128/JB.01044-10
Jaffe AB, Hall A (2005) Rho GTPases: biochemistry and biology. Annu Rev Cell Dev Biol 21:247–269
Lacey LA, Chauvin RL (1999) Entomopathogenic nematodes for control of diapausing codling moth (Lepidoptera: Tortricidae) in fruit bins. J Econ Entomol 92(1):104–109
Lacey LA, Georgis R (2012) Entomopathogenic nematodes for control of insect pests above and below ground with comments on commercial production. J Nematol 44(2):218–225
Landsberg MJ, Jones SA, Rothnagel R, Busby JN, Marshall SD, Simpson RM, Lott JS, Hankamer B, Hurst MR (2011) 3D structure of the *Yersinia entomophaga* toxin complex and implications for insecticidal activity. Proc Natl Acad Sci USA 108(51):20544–20549. doi: 10.1073/pnas.1111155108, [pii] 1111155108
Lang AE, Konukiewitz J, Aktories K, Benz R (2013) TcdA1 of *Photorhabdus luminescens*: Electrophysiological Analysis of Pore Formation and Effector Binding. Biophys J 105(2):376–384. doi: 10.1016/j.bpj.2013.06.003, [pii] S0006-3495(13)00675-9
Lang AE, Qu Z, Schwan C, Silvan U, Unger A, Schoenenberger CA, Aktories K, Mannherz HG (2016) Actin ADP-ribosylation at Threonine148 by *Photorhabdus luminescens* toxin TccC3 induces aggregation of intracellular F-actin. Cell Microbiol. doi:10.1111/cmi.12636
Lang AE, Schmidt G, Schlosser A, Hey TD, Larrinua IM, Sheets JJ, Mannherz HG, Aktories K (2010) *Photorhabdus luminescens* toxins ADP-ribosylate actin and RhoA to force actin clustering. Science 327(5969):1139–1142. doi: 10.1126/science.1184557, [pii] 327/5969/1139
Lee SC, Stoilova-McPhie S, Baxter L, Fulop V, Henderson J, Rodger A, Roper DI, Scott DJ, Smith CJ, Morgan JA (2007) Structural characterisation of the insecticidal toxin XptA1, reveals a 1.15 MDa tetramer with a cage-like structure. J Mol Biol 366(5):1558–1568. doi: 10.1016/j.jmb.2006.12.057, [pii] S0022-2836(06)01747-5
Lemichez E, Aktories K (2013) Hijacking of Rho GTPases during bacterial infection. Exp Cell Res 319(15):2329–2336
Liu D, Burton S, Glancy T, Li ZS, Hampton R, Meade T, Merlo DJ (2003) Insect resistance conferred by 283-kDa *Photorhabdus luminescens* protein TcdA in *Arabidopsis thaliana*. Nat Biotechnol 21(10):1222–1228
Mannherz HG, Hannappel E (2009) The beta-thymosins: intracellular and extracellular activities of a versatile actin binding protein family. Cell Motil Cytoskeleton 66(10):839–851. doi:10.1002/cm.20371[doi]
Medjkane S, Perez-Sanchez C, Gaggioli C, Sahai E, Treisman R (2009) Myocardin-related transcription factors and SRF are required for cytoskeletal dynamics and experimental metastasis. Nat Cell Biol 11(3):257–268. doi: 10.1038/ncb1833, [pii] ncb1833
Meusch D, Gatsogiannis C, Efremov RG, Lang AE, Hofnagel O, Vetter IR, Aktories K, Raunser S (2014) Mechanism of Tc toxin action revealed in molecular detail. Nature 508(7494):61–65. doi: 10.1038/nature13015, [pii] nature13015
Morgan JA, Sergeant M, Ellis D, Ousley M, Jarrett P (2001) Sequence analysis of insecticidal genes from *Xenorhabdus nematophilus* PMFI296. Appl Environ Microbiol 67(5):2062–2069. doi:10.1128/AEM.67.5.2062-2069.2001

Parkhill J, Dougan G, James KD, Thomson NR, Pickard D, Wain J, Churcher C, Mungall KL, Bentley SD, Holden MT, Sebaihia M, Baker S, Basham D, Brooks K, Chillingworth T, Connerton P, Cronin A, Davis P, Davies RM, Dowd L, White N, Farrar J, Feltwell T, Hamlin N, Haque A, Hien TT, Holroyd S, Jagels K, Krogh A, Larsen TS, Leather S, Moule S, O'Gaora P, Parry C, Quail M, Rutherford K, Simmonds M, Skelton J, Stevens K, Whitehead S, Barrell BG (2001) Complete genome sequence of a multiple drug resistant *Salmonella enterica* serovar Typhi CT18. Nature 413(6858):848–852

Pfaumann V, Lang AE, Schwan C, Schmidt G, Aktories K (2015) The actin and Rho-modifying toxins PTC3 and PTC5 of *Photorhabdus luminescens*: enzyme characterization and induction of MAL/SRF-dependent transcription. Cell Microbiol 17(4):579–594. doi:10.1111/cmi.12386 [doi]

Pinto AF, Schuler H (2015) Comparative structural analysis of the putative mono-ADP-ribosyltransferases of the ARTD/PARP family. Curr Top Microbiol Immunol 384:153–166. doi:10.1007/82_2014_417

Pollard TD, Cooper JA (2009) Actin, a central player in cell shape and movement. Science 326 (5957):1208–1212. doi: 10.1126/science.1175862, [pii] 326/5957/1208

Posern G, Treisman R (2006) Actin' together: serum response factor, its cofactors and the link to signal transduction. Trends Cell Biol 16 (11):588–596. doi: 10.1016/j.tcb.2006.09.008, [pii] S0962-8924(06)00269-8

Scheffzek K, Ahmadian MR, Wittinghofer A (1998) GTPase-activating proteins: helping hands to complement an active site. Trends Biochem Sci 23(7):257–262. [pii] S0968-0004(98)01224-9

Schering B, Bärmann M, Chhatwal GS, Geipel U, Aktories K (1988) ADP-ribosylation of skeletal muscle and non- muscle actin by *Clostridium perfringens* iota toxin. Eur J Biochem 171:225–229

Schmidt G, Sehr P, Wilm M, Selzer J, Mann M, Aktories K (1997) Gln63 of Rho is deamidated by *Escherichia coli* cytotoxic necrotizing factor 1. Nature 387:725–729

Schwan C, Stecher B, Tzivelekidis T, van HM, Rohde M, Hardt WD, Wehland J, Aktories K (2009) *Clostridium difficile* toxin CDT induces formation of microtubule-based protrusions and increases adherence of bacteria. PLoS Pathog 5(10):e1000626. doi:10.1371/journal.ppat.1000626 [doi]

Sergeant M, Baxter L, Jarrett P, Shaw E, Ousley M, Winstanley C, Morgan JA (2006) Identification, typing, and insecticidal activity of *Xenorhabdus* isolates from entomopathogenic nematodes in United Kingdom soil and characterization of the xpt toxin loci. Appl Environ Microbiol 72(9):5895–5907. doi:10.1128/AEM.00217-06

Sergeant M, Jarrett P, Ousley M, Morgan JA (2003) Interactions of insecticidal toxin gene products from *Xenorhabdus nematophilus* PMFI296. Appl Environ Microbiol 69(6):3344–3349

Sheets JJ, Hey TD, Fencil KJ, Burton SL, Ni W, Lang AE, Benz R, Aktories K (2011) Insecticidal toxin complex proteins from *Xenorhabdus nematophilus*: structure and pore formation. J Biol Chem 286(26):22742–22749. doi: 10.1074/jbc.M111.227009, [pii] M111.227009

Silva CP, Waterfield NR, Daborn PJ, Dean P, Chilver T, Au CP, Sharma S, Potter U, Reynolds SE (2002) Bacterial infection of a model insect: *Photorhabdus luminescens* and *Manduca sexta*. Cell Microbiol 4(6):329–339. doi:194 [pii]

Simon NC, Aktories K, Barbieri JT (2014) Novel bacterial ADP-ribosylating toxins: structure and function. Nat Rev Microbiol 12(9):599–611. doi: 10.1038/nrmicro3310, [pii] nrmicro3310

Tezcan-Merdol D, Nyman T, Lindberg U, Haag F, Koch-Nolte F, Rhen M (2001) Actin is ADP-ribosylated by the *Salmonella enterica* virulence-associated protein SpvB. Mol Microbio 39:606–619

Thumkeo D, Watanabe S, Narumiya S (2013) Physiological roles of Rho and Rho effectors in mammals. Eur J Cell Biol 92(10–11):303–315. doi: 10.1016/j.ejcb.2013.09.002, [pii] S0171-9335(13)00058-7

Vandekerckhove J, Schering B, Bärmann M, Aktories K (1987) *Clostridium perfringens* iota toxin ADP-ribosylates skeletal muscle actin in Arg-177. FEBS Lett 225:48–52

Vandekerckhove J, Schering B, Bärmann M, Aktories K (1988) Botulinum C2 toxin ADP-ribosylates cytoplasmic β/g-actin in arginine 177. J Biol Chem 263:696–700

Vilches S, Wilhelms M, Yu HB, Leung KY, Tomas JM, Merino S (2008) *Aeromonas hydrophila* AH-3 AexT is an ADP-ribosylating toxin secreted through the type III secretion system. Microb Pathog 44(1):1–12. doi: 10.1016/j.micpath.2007.06.004, [pii] S0882-4010(07)00084-8

Visschedyk DD, Perieteanu AA, Turgeon ZJ, Fieldhouse RJ, Dawson JF, Merrill AR (2010) Photox, a novel actin-targeting mono-ADP-ribosyltransferase from *Photorhabdus luminescens*. J Biol Chem 285(18):13525–13534. doi: 10.1074/jbc.M109.077339, [pii] M109.077339

Vogelsgesang M, Pautsch A, Aktories K (2007) C3 exoenzymes, novel insights into structure and action of Rho-ADP-ribosylating toxins. Naunyn Schmiedebergs Arch Pharmacol 374(5–6):347–360

Waterfield N, Dowling A, Sharma S, Daborn PJ, Potter U, ffrench-Constant RH (2001a) Oral toxicity of *Photorhabdus luminescens* W14 toxin complexes in Escherichia coli. Appl Environ Microbiol 67(11):5017–5024

Waterfield N, Hares M, Yang G, Dowling A, ffrench-Constant R (2005) Potentiation and cellular phenotypes of the insecticidal Toxin complexes of *Photorhabdus* bacteria. Cell Microbiol 7 (3):373–382

Waterfield NR, Bowen DJ, Fetherston JD, Perry RD, ffrench-Constant RH (2001b) The tc genes of *Photorhabdus*: a growing family. Trends Microbiol 9(4):185–191

Wegner A, Aktories K (1988) ADP-ribosylated actin caps the barbed ends of actin filaments. J Biol Chem 263:13739–13742

Wilkinson P, Waterfield NR, Crossman L, Corton C, Sanchez-Contreras M, Vlisidou I, Barron A, Bignell A, Clark L, Ormond D, Mayho M, Bason N, Smith F, Simmonds M, Churcher C, Harris D, Thompson NR, Quail M, Parkhill J, ffrench-Constant RH (2009) Comparative genomics of the emerging human pathogen *Photorhabdus asymbiotica* with the insect pathogen *Photorhabdus luminescens*. BMC Genomics 10:302. doi: 10.1186/1471-2164-10-302, [pii] 1471-2164-10-302

Yang G, Waterfield NR (2013) The role of TcdB and TccC subunits in secretion of the *Photorhabdus* Tcd toxin complex. PLoS Pathog 9(10):e1003644. doi:10.1371/journal.ppat.1003644

Identifying Anti-host Effectors in *Photorhabdus*

Andrea J. Dowling

Abstract The death of the insect host is an essential part of the life cycle of *Photorhabdus,* and as a result, this bacterium comes equipped with a dazzlingly large array of toxins and virulence factors that ensure rapid insect death. Elucidation of the key players in insect infection and mortality has therefore proved difficult using traditional microbiological techniques such as individual gene knockouts due to the high level of functional redundancy displayed by *Photorhabdus* virulence factors. Thus, knockout of any individual toxin gene may serve to delay time to death but not to render the bacteria avirulent due to the continued presence of an array of other toxins and virulence factors in the single-gene mutant. This functional redundancy had led to the necessary development of an array of techniques and new model systems for identifying and dissecting apart the action of anti-insect effectors produced by *Photorhabdus*. These have been pivotal in both the identification of new toxins and virulence factors and in ascribing functions to them. These techniques have gone on to prove valuable in pathogenic bacteria other than *Photorhabdus* and are likely to be useful in many others.

Contents

A.J. Dowling (✉)
Biosciences, College of Life and Environmental Sciences, University of Exeter, Penryn Campus, Cornwall TR10 9FE, UK
e-mail: A.J.Dowling@exeter.ac.uk

Current Topics in Microbiology and Immunology (2017) 402:25–38
DOI 10.1007/82_2016_51

Published Online: 26 December 2016

1 Introduction

Photorhabdus bacteria have a complex multi-host lifestyle in which they must switch from being a nematode mutualist to a potent insect pathogen. Insect death is a critical part of the life cycle and as such *Photorhabdus* produces a staggering array of anti-host effectors, those characterized so far include toxins, adhesins, enzymes, secondary metabolites and secretion systems amongst others (see also Chapter by Hapeshi and Waterfield in this volume). The discovery and elucidation of the anti-host effectors are not only important in understanding how *Photorhabdus* bacteria may interact with and infect insect and mammalian hosts but it also makes them an attractive source for novel insecticides and drugs. However, the discovery and understanding of such anti-host effectors via classical microbiology (single-gene knockouts) are exacerbated due to the issue of functional redundancy. Mutating a single toxin gene in *Photorhabdus* often results in very little or indeed no attenuation in pathogenicity, as there are a host of other toxins and effectors that compensate and insure death of the host following infection with a single-gene knockout. This has led to the development of several novel and innovative screening assays designed to circumvent functional redundancy and identify anti-host effectors on a genome-wide scale in vivo. A number of bioinformatic-based *in silico* techniques have also identified toxins and novel secretions systems from *Photorhabdus* that have biological activity. This chapter will review these approaches and discuss future directions for this research.

2 A Forward Genetics Approach

Single-gene knockouts in predicted toxins are often ineffectual in significantly reducing the pathogenicity of *Photorhabdus* due to a high level of functional redundancy. In order to circumvent this problem and elucidate the masked action of individual anti-insect effectors, a simple 'forward genetic' (gain of function) approach and the use of a model insect system for screening were developed. This original approach initially involved generating a cosmid library of *P. luminescens* subsp. *akhurstii* strain W14 in *Escherichia coli* with each cosmid clone in *E. coli* containing an insert of $\sim$40 kb of *Photorhabdus* genomic DNA. Cosmids were arrayed into 96-well micro-plates in order to give $\sim$10X genomic coverage. Single

E. coli clones carrying individual cosmids were then injected separately into caterpillars of the tobacco hornworm, *Manduca sexta*. Laboratory strain *E. coli* is normally cleared successfully by the *M. sexta* immune system; however, one in every 300 of the cosmids injected caused the recipient caterpillar to lose turgor 12 h post-injection and subsequent insect death occurred at 24 h (Daborn et al. 2002). Insertional (transposon) mutagenesis of the associated cosmid revealed that a single 8.8 kb gene was responsible for this phenotype that was then termed *makes caterpillars floppy* or the *mcf* gene. The encoded Mcf protein toxin (324 KDa) facilitates the persistence of *E. coli* in the insect haemocoel and causes massive cell death via apoptosis of both haemocytes and midgut cells resulting in insect mortality.

Random end sequencing of the *P. luminescens* W14 cosmid library used in this screen identified a cosmid containing sequence which assembled to that of *mcf* but displayed clear differences at the nucleotide level. This newly identified homologue was named *makes caterpillars floppy 2* or *mcf2,* and correspondingly, *mcf* was renamed *mcf1*. *E. coli* expressing the Mcf2 toxin display similar insect toxicity to Lepidopteran larvae as Mcf1 (Waterfield et al. 2003).

The *mcf2* gene predicts another large (262 kDa) protein, and like Mcf1, amino acids 1015–1548 of Mcf2 are 39% similar and 20% identical to amino acids 867–1368 of *C. difficile* toxin B, but Mcf2 lacks the similarity to the C-terminus of an RTX-like toxin from *Actinobacillus pleuropneumoniae* carried by the C-terminus of Mcf1. Since the original description of the Mcf1 gene, a new entry in GenBank, shows that both Mcf1 and Mcf2 also show a second region of homology to an RTX-like cytotoxin (gene VVA1030) from *Vibrio vulnificus* strain YJ016, amino acids 695–914 of Mcf2 are 30% identical and 48% similar to amino acids 3250–3496 of VVA1030. The predicted N-terminal domain of Mcf2 is shorter than that of Mcf1 and lacks a BH3-like domain. Instead, the N-terminus carries a region with significant similarity (54% similarity and 40% identity) to the C-terminus of the plant avirulence protein HrmA from *P. syringae* pv. *syringae*. Further examination of unfinished genome sequences shows that Mcf1 and Mcf2 are part of a family of toxins including two undescribed homologues from *Pseudomonas fluorescens* strain Pf01, ORFs 4315 and 4316, which are 28% identical and 46% similar to each other. Domains containing *mcf* sequence are also present in the MARTX toxin family (Fullner Satchell 2007; Roig et al. 2011).

2.1 *Rapid Virulence Annotation*

One species of *Photorhabdus*, termed *P. asymbiotica,* is a pathogen of both insects and mammals. Originally isolated from a human infection, and with no nematode present, it was initially presumed to be literally 'asymbiotic', hence the name '*asymbiotica*'. However, a nematode vector for *P. asymbiotica* has since been found (Gerrard et al. 2006). For a detailed description of *P. asymbiotica,* see chapters by John Gerrard also in this volume. Analysis of *P. asymbiotica* therefore provides a unique opportunity to identify and explore virulence factors which are specific for either insect or mammalian hosts, or indeed active against both (see also Chapter by Hapeshi and Waterfield in this volume).

The success of the forward genetic approach in revealing toxins otherwise disguised due to functional redundancy led to the pioneering Rapid Virulence Annotation (RVA) approach developed by Waterfield et al. (2008). In this approach, parallel assays were used to find effectors in a cosmid library of *P. asymbiotica* ATCC43949 with toxicity against a range of taxa: insects (*M. sexta* and *Galleria mellonella*), amoebae (*Acanthamoeba polyphaga*), nematodes (*Caenorhabditis elegans*) and mouse macrophages (J774-2 murine cell line). This approach allows us to detect effectors with species specific or broad-spectrum activity of individual clones that confer a gain of toxicity (GOT) to the library strain *E. coli* expressing them. Assembling the end sequences of clones that confer toxicity onto the *P. asymbiotica* genome is used to identify the virulence-related regions. The multiplicity of genome coverage within the library ($\sim$10X) means that when the end sequences of the positive cosmids are aligned, they generate overlapping clusters (minimum of two overlapping cosmids) allowing the identification of the minimum genomic region of interest (ROI) carrying the candidate effector ORFs denoted as RVA regions (Fig. 1). Twenty-one RVA regions were identified in the *P. asymbiotica* genome, and these included the re-identification of known virulence factors such as the toxin Mcf1. These assays revealed activity of Mcf1 against both *C. elegans* and amoebae, in addition to its known activity against insects and mammalian tissue culture cells, indicating a broad mechanism of action (Daborn et al. 2002; Dowling et al. 2004).

The other twenty RVA regions contained gene clusters and predicted functions related to: Type III secretion, hemolysins, hemagglutinins, *Photorhabdus* Virulence Cassettes (PVCs), production of secondary metabolite small molecules (non-ribosomal peptide and polyketide synthetases), Type VI secretion, lipases, enterotoxins, fimbrial operons, rtxA toxin homologues and several containing genes of unknown function.

2.2 Secondary Metabolites as Virulence Factors

Photorhabdus contains a large number of gene clusters associated with the synthesis of secondary metabolites, the functions of many of which are unknown. A third of the RVA clusters identified contain polyketide synthetases (PKS) and/ or non-ribosomal peptide synthases (NRPS) gene clusters, including a yersiniabactin-like compound, a virulence factor of *Yersinia pestis* causative agent of the black plague (Bearden et al. 1997). The large number of RVA regions carrying these clusters provides a strong link between certain secondary metabolites and a previously unknown role in *Photorhabdus* virulence. For more detailed references on secondary metabolites, see also chapters by Helge Bode and David Clarke in this volume.

2.3 Host Specificity

Several regions encoding putative lipase genes (*pdl*) were also identified in these RVA screens. These regions were active against insects and nematodes but not on murine macrophages, suggestive of host specificity. However, the majority of RVA regions active against macrophages were also active against insects indicating that factors active against the insect are also capable of effect on mammalian hosts. Fine scale mapping using insertion (transposon) mutagenesis is used on positive cosmids to directly identify the genes responsible for the phenotype. The mutants can then be rescreened in the desired assay to look for clones displaying a loss of function. These clones can then be sequenced in order to identify the gene/s responsible for the activity. Cloning the candidate gene into *E. coli* and retesting this clone in the requisite assay are then carried out in order to confirm activity.

The unbiased nature of this approach allows for the identification of genes that would not otherwise be identified as virulence factors. Insertional mutagenesis of cosmids from RVA17 that were toxic to insects, nematodes and macrophages revealed that virulence was linked to expression of a *kdp* operon encoding a Kdp potassium pump. Kdp sytems are involved in potassium homeostasis perhaps making this an unexpected candidate virulence factor. However, follow-up studies have revealed that this KdpD/KdpE two-component system enables library strain *E. coli to* persist within insect haemocytes, eventually leading to mortality and causes a dramatic increase in phagocytic activity in murine macrophages (Vlisidou et al. 2010; Dowling and Hodgson 2014).

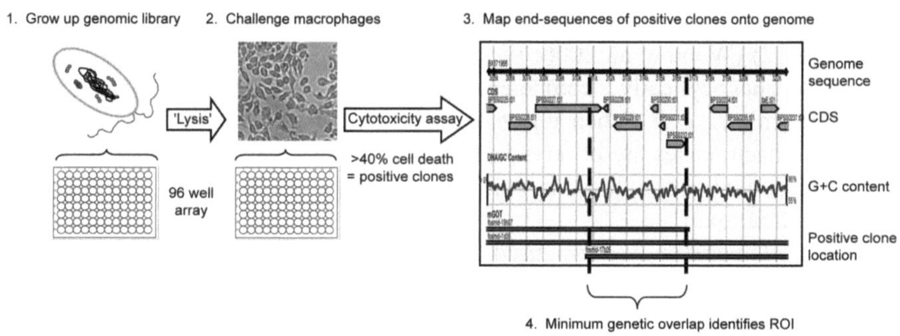

Fig. 1 Workflow of a typical RVA-type screen for anti-phagocyte effectors. A genomic library of the strain of *Photorhabdus* of interest is made in cosmids or fosmids and transformed into library strain *E. coli*. Individual clones are arrayed in a 96-well plate format to achieve ~ 10X coverage of the genome and are then end-sequenced. Library plates are grown up, and crude lysate preparations of these are applied to phagocytes (macrophages or haemocytes) similarly arrayed in a 96-well plate. These are co-incubated for the desired length of time before assaying for cytotoxicity. Clones causing >40% cell death are called as 'positive', and the end sequences for these are assembled onto the genome sequence of the relevant *Photorhabdus* strain. Any genomic region with two or more overlapping clones mapped across it is designated a region of interest for further follow-up

This approach has further untapped potential, and the capacity for detection of effectors can easily be expanded with screens and assays designed to look for effectors, not simply involved in toxicity, but those involved in other pathogenicity traits such as adherence, invasion and intracellular survival.

3 The *Drosophila* Embryo System

Once an effector is identified, it is then necessary to establish its function and mode of action in infection. *Drosophila* are a well-established powerful genetic model organism used to study microbial infections. In vivo studies are largely based on end-point analysis and a system in which infection can be studied in real time is therefore extremely useful for a continuous view of infection. The *Drosophila* embryo system, previously established as a model for studying development and wound repair, has also been developed as a tool which allows visualization of the early stages of infection and toxin action in real time using confocal time-lapse microscopy (Wood et al. 2006; Vlisidou et al. 2009). By using a combination of *Drosophila* genetics to look at different host genetic backgrounds and time-lapse confocal microscopy to follow bacterial infection in real time, it is possible to see how different mutant bacteria (or recombinant *E. coli* armed with *Photorhabdus* toxins) perform against different host genetic backgrounds. Micro-injection is used to introduce a fluorescently labelled microbe of interest, or purified toxin, into a recombinant *Drosophila* embryo with haemocytes expressing a fluorescent protein such as GFP. Embryonic haemocytes are motile macrophage-like cells that migrate throughout the developing embryo following stereotypical routes to disperse from their point of origin to eventually distribute themselves equally throughout the insect (Fig. 2) (Wood et al. 2006; Wood and Jacinto 2007). These haemocytes are dynamic, phagocytically competent and able to recognize and engulf non-pathogenic *E. coli*. When injected with the *P. asymbiotica* ATCC43949, the embryonic haemocytes are observed to undergo rapid paralysis or 'freezing'. This phenotype could be reproduced by injection *E. coli* expressing Mcf1 or by injection of the purified toxin.

Drosophila genetics was then used to screen for mechanism of action. Mcf1 had previously been identified as requiring internalization via endocytosis in order to exert its pro-apoptotic activity (Dowling et al. 2007), and by using a mutant deficient in (clathrin-dependent) endocytic machinery (*shibire*^{ts1} mutant), we were able to confirm that these haemocytes were unaffected by Mcf1 and did not freeze, meaning that internalization was also necessary for this phenotype.

The rapid effect on the hemocyte cytoskeleton led to the hypothesis that Mcf1 may be acting on a pre-existing eukaryotic molecular switch involved in cytoskeletal dynamics within the cell such as the Rho GTPases. Micro-injection of *Drosophila* mutants expressing dominant negative or constitutively active small GTPase Rac revealed that these haemocytes also failed to freeze in response to Mcf1. This indicates a need for the presence of wild-type Rac in order for the

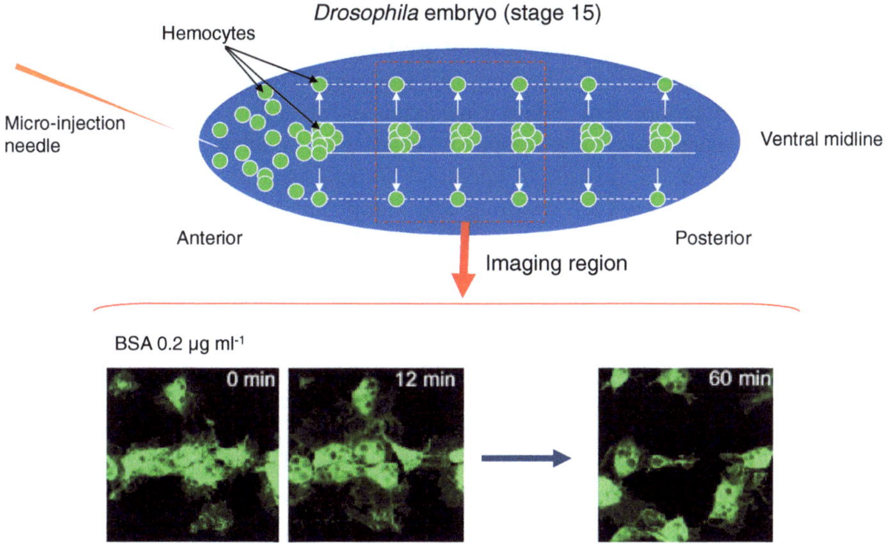

Fig. 2 The *Drosophila* embryo system. Stage 15 *Drosophila* embryos with haemocytes expressing a fluorescent protein such as GFP are immobilized on a microscope slide prior to anterior micro-injection with the bacteria or purified protein of interest. Z-stack time-lapse confocal microscopy is used to image the activity of the haemocytes post-injection in order to monitor their subsequent behaviour phagocytosis and observe the progress of the infection in real time

freezing process to occur. Several pathogenic bacteria produce toxins that inactivate Rho GTPases in order to avoid phagocytosis, known as RhoGAPs (Aktories 2011; Popoff 2014). However, whether Mcf1 is acting as a RhoGAP remains to be confirmed. This screening method is a powerful tool for identifying effector mode of action in vivo, and *Drosophila* genetic mutants present a valuable method of clearly dissecting many aspects of function in order to elucidate toxin mode of action and likely role in infection. Using *Drosophila* genetic mutants, we were able to determine that Mcf1 required internalization in order to exert activity and that the freezing phenotype was likely caused by Mcf1 acting on Rac.

4 Immune Cell Phenotype Screen

Although a cytotoxicity screen can identify effectors, it does not point us in the direction of how it is exerting effect on host cells. Typically visual analysis methods, such as fluorescence microscopy, are used to explore bacterial effector interaction with host cells. The original RVA screen for anti-macrophage factors from *P. asymbiotica* produced a wealth of candidates requiring further investigation to identify function (23 regions of interest covered by over 80 cosmids). Follow-up of this number of putative effectors would be very labour-intensive and time-consuming to follow up on a case by case basis. In order to address this, we developed a novel genome scale morphology-based analysis method to functionally group the bacterial effectors (Dowling and Hodgson 2014).

Numerous functions and putative mechanisms of action can be characterized based on cell morphology, e.g. apoptosis (cellular and nuclear shrinkage followed by fragmentation and formation of apoptotic bodies), necrosis (cell swelling, leaky membranes), pyroptosis (cell swelling and significant cell size increase, rupture, nuclear condensation but integrity remains) and alteration in actin cytoskeleton, such as formation of stress fibres or depolymerisation, indicating activity on Rho GTPases (Gruenheid and Finlay 2003; Duprez et al. 2009; Lamkanfi and Dixit 2010; Aktories 2011). The targets of many bacterial pathogens effectors are the actin cytoskeleton and the nuclei where they manipulate or usurp normal cellular function or induce cell death to their advantage in establishing infection. The analysis uses high-content image analysis to visualize and quantify the morphological alterations induced in macrophages in response to treatment with preparations from the cosmid library clones. Phenotypes induced by the individual clones were quantified by measuring the morphology, staining intensity, and spatial attributes of the cellular cytoskeleton and nuclei. Application of statistical multivariate analysis to multiple cellular measures obtained from the treated macrophages revealed that we could group macrophage response based on phenotype. Five significantly distinct phenotype clusters linked to effector function were discovered (Fig. 3). Two of the clusters described control library *E. coli* lysate-treated

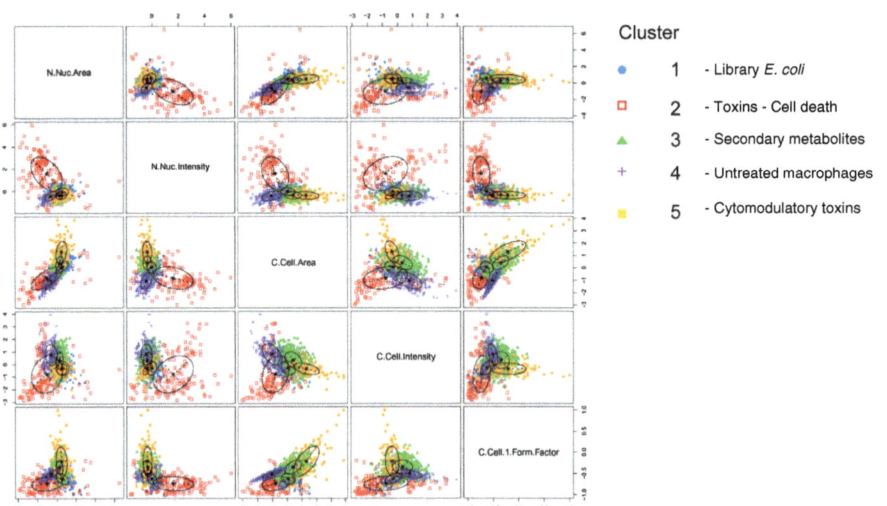

Fig. 3 Typical output from a multivariate cluster analysis screen. Combining high-throughput high-content analysis microscopy as typically employed in pharmaceutical drug discovery. Plots are being made using morphometric measures in order to reveal and group distinct phenotypic responses from cells treated with individual *Photorhabdus* cosmid library clones. Here, five clusters in phenotype space are based on measures of two nuclear parameters: area and intensity and three cell parameters: area, intensity and form factor

macrophages and untreated macrophages, representing the normal cellular morphology. Three of the clusters describe alterations in cell morphology that link effectors with putative function. Cluster 2 describes nuclear and cytoplasmic condensation and collapse of the actin cytoskeleton. This cluster groups clones from regions of interest that include known toxins: Mcf1, toxin complexes (Tcs), XaxAB, along with putative RtxA-like toxins and Type VI secretion system amongst others. The grouping of Mcf1 into this cluster acts here as a 'phenotypic hallmark' indicating that the other candidate effectors in this group are likely to be highly cytotoxic, possibly pro-apoptotic toxins also. The cellular characteristics induced by effectors described by Cluster 3 are increased nuclear and cell size. The dominant predicted function of candidate effectors grouped into this cluster is linked to bacterial adhesion to the host cell and intracellular survival, including adhesins and hemagglutinins. However, Cluster 3 also contains candidates associated with synthesis of secondary metabolites, indicating that small molecules may play an important role in usurping macrophage function during *Photorhabdus* infection.

Interestingly, during the infection cycle of *P. asymbiotica,* both evasion of phagocytosis and intracellular survival play an important role (Costa et al. 2010). In order to ascribe phagocytic function to phenotype cluster, an additional screen was developed to detect alterations in macrophage phagocytosis in response to treatment with positive library clone preparations. Importantly, this screen found that pro-phagocytic and anti-phagocytic effectors belong to different clusters. Pro-phagocytic candidates belong exclusively to Cluster 3, and anti-phagocytic candidates are found in clusters 2 and 5. The most significant decrease in phagocytosis was caused by treatment with preparations containing PNF (*Photorhabdus* necrosis factor) (ROI 19, Cluster 5) and Mcf1 (ROI 20, Cluster 2). The most significant increase in phagocytosis is linked to ROI 14 that encodes a Kdp potassium transport system and a MACPF-domain perforin-like protein. As mentioned previously, the Kdp system from *P. asymbiotica* has been linked with intracellular survival in haemocytes (Vlisidou et al. 2010).

Potassium is considered to be one of the key regulators of cell response to pore formation. The cytosolic drop in K+ normally associated with pore forming is associated with a number of different effects: autophagy, translation arrest, and the activation of MAPK pathways and proteolytic cascades (Bischofberger et al. 2012). Potassium is also important in the maturation of the phagolysosome important for killing engulfed microbes. The effectors contained within this ROI may play a role in the intracellular life stage of *P. asymbiotica*, but this is yet to be confirmed.

However, the mechanism by which the Kdp system and/or perforin is facilitating intracellular survival or promoting phagocytosis remains unclear.

Intriguingly, several cosmids were identified that contained *mcf1* and a tightly linked downstream NRPS, whereas others contained only *mcf1* or the downstream NRPS, indicating that NRPS gene product has a cytotoxic activity of its own. Treatment of macrophages with a combination of lysate preparations from two cosmids containing *mcf1* alone or NRPS alone caused increased cytotoxicity and dramatic cytoskeletal alterations compared to the activity of lysates from cosmids carrying just one or the other. Further, injection of cosmids expressing both into *Galleria mellonella* larvae caused higher mortality that injection of cosmids carrying either mcf1 or the NRPS alone.

5 *In Silico* Screens

Bioinformatic tools and motif searching have also been used successfully to reveal novel toxins in *Photorhabdus* taking advantage of the complete genome sequences available for several strains.

5.1 *Photox—A Toxin Unique to* P. luminescens

A novel mono-ADP-ribosyltransferase targeting toxin or 'mART' toxin was identified from *Photorhabdus luminescens* TT01 using a bioinformatic approach involving searching the genomic threading database using a shared core structure sequence identity (SCOP code) designed to identify new putative mART toxins (Visschedyk et al. 2010). As sequence identity between mARTs is low, the technique looks at both the primary amino acid sequence in order to identify key catalytic regions and the predicted fold pattern. A gene encoding a predicted protein of 45.9 kDa, *plu0822*, with significant identity to known mART toxins was discovered. Purified Photox possesses relatively high ADP-ribosyltransferase activity, and it specifically targets Arg^{177} of actin preventing actin polymerization. Expression of Photox in yeast cells caused a severe defect in growth indicating high toxicity of the protein. Photox has a predicted 185 aa 'disordered' N-terminal region of unknown function, with no homologues identified using BLAST. It is hypothesized that this N-terminal region may potentially be involved in entry into the host cell. However, the precise role of this toxin during *Photorhabdus* infection currently remains unclear.

5.2 *PaTox—A Toxin Unique to* P. asymbiotica

Bio-active domain-focused bioinformatic analysis has been used successfully in several cases to identify novel *Photorhabdus* effectors. The *P. asymbiotica* toxin PaTox was discovered by using basic local alignment search tool (BLAST) analysis to look for sequence motifs (DxD) from known glycosylating toxins within the genome (Jank et al. 2013). A gene PAU_02230 unique to *P. asymbiotica* was found encoding a putative protein toxin of $\sim$3 kb with a potential C-terminal glycosyltransferase domain, and further analysis revealed a downstream region with 68% sequence similarity with the *Salmonella* virulence factor Sse1, whose function is unknown. The putative toxin was named *P. asymbiotica* toxin 'PaTox'. Injection of full-length recombinant purified PaTox protein into *G. mellonella* larvae resulted in 100% insect mortality within ~3.5 days. Disruption of the DxD motif resulted in >60% of survival. Further, delivery of the glycosyltransferase domain into J774 murine macrophages and HeLa cells using anthrax protective antigen (PA) resulted in phagocytic block and breakdown of the actin cytoskeleton. Again, disruption of the DxD motif attenuated these activities proving the essential role of this domain in cytotoxicity. PaTox targets the small GTPase Rho, a target for several known bacterial effectors and toxins. Additionally, the SseI-like domain deamidates heterotrimeric G proteins as discovered upon testing a PaTox mutant deficient in DxD but containing SseI on HeLa cells the significant formation of stress fibres was observed indicating activation of RhoA. Therefore, PaTox is an AB-like toxin with two C-terminal catalytic domains and an N-terminal domain that promotes receptor translocation and enters host cells via endosomes. Indeed, tyrosine GlcNAcylation is detected during *P. asymbiotica* infection of insects but not during *P. luminescens* infection; PaTox is unique to *P. asymbiotica* and is likely responsible for this phenotype. However, again the exact role of this toxin in the infection process in human/ mammalian infection has yet to be elucidated.

5.3 The Photorhabdus *Virulence Cassettes*

Analysis of the complete genome sequences of *P. asymbiotica* ATCC43949 and *P. luminescens* TT01 revealed a noticeably significant number of prophage-like loci throughout the genomes of both species. These were initially recognized due to their similarity to a prophage-like locus in the pADAP plasmid from the insect pathogen *Serratia entomophila*. The pADAP plasmid is responsible for causing 'amber disease' in New Zealand grass grubs caused by homologues of the *tcs* (*sepA*, *B* and *C*); alongside this, the prophage-like locus is responsible for a separate anti-feeding effect highlighting such prophage-like loci as playing a role in pathogenesis (Hurst et al. 2004). The pADAP prophage-like locus is divided into two regions, one contains 18 putative ORFs predicting high similarity to phage tail and base plate proteins linked to a second containing a predicted putative effector protein thought to be responsible for the anti-feeding phenotype. The prophage-like loci from *Photorhabdus* were identified as they have the same genomic organization as the pADAP prophage-like locus, with a conserved phage-like unit but with numerous different putative effector domains, termed the *Photorhabdus* Virulence Cassettes (PVCs) (Yang et al. 2006). Analysis of the putative effectors revealed similarity to sections of known toxins including Mcf1, the active site of cytotoxic necrosis factor CNF1 from *E. coli* (Pnf) amongst others; further, several have no similarity to known toxins and possibly represent novel effectors. In order to test, toxicity towards insects was tested by injecting *E. coli* carrying PVC-expressing cosmid clones into *Galleria mellonella*. All PVCs from *P. asymbiotica* were found to cause high levels of mortality, whereas intriguingly mortality was much less pronounced in those injected with *E. coli* expressing PVC carrying cosmids from *P. luminescens* (Yang et al. 2006). The PVCs represent a very interesting novel group of effector delivery systems with much to be elucidated about how they secrete and target and their effectors into host cells.

6 Future Approaches

In the quest to understand both *Photorhabdus* biology and discover novel anti-host effectors with the potential for use in agriculture, there are several technological advances and experimental avenues of particular interest. The variety of screening efforts to date has identified a wealth of effector proteins and molecules from *Photorhabdus*, the detailed activity of many of which remain to be elucidated. Transcriptomic analysis of *Photorhabdus* gene expression using RNA-Seq represents a very promising approach for understanding gene expression in response to certain environmental conditions, and some headway has been made using this approach in vitro (Mulley et al. 2015). Looking at gene expression in vivo as a function of host infection will be a valuable tool in understanding not only the anti-host effectors involved in usurping insect immunity, but also when they are deployed over the course of infection.

Finally, target host, infection biology or mode-of-action-focused extensions of library screens would prove a valuable opportunity to both discover further effectors and also understand how they are acting on hosts and host cells. Library assays could focus on activity against different insect hosts or specificity towards different tissues, e.g. haemocytes or midgut, or on conferring the gain or loss of different virulence phenotypes onto library *E. coli* towards mammalian or insect host cells. Examples of such assays could include the following: gain of adhesion, gain of intracellular survival and persistence and gain or loss of phagocytosis. More 'mode-of-action'-based approaches could look at mechanisms of cell death (e.g. apoptosis, necrosis, pyroptosis), alterations to the cell cycle or effects on subcellular targets including the actin cytoskeleton, Golgi apparatus and mitochondria.

References

Aktories K (2011) Bacterial protein toxins that modify host regulatory GTPases. Nat Rev Microbiol 9:487–498. doi:10.1038/nrmicro2592

Bearden SW, Fetherston JD, Perry RD (1997) Genetic organization of the yersiniabactin biosynthetic region and construction of avirulent mutants in *Yersinia pestis*. Infect Immun 65:1659–1668

Bischofberger M, Iacovache I, van der Goot FG (2012) Pathogenic pore-forming proteins: function and host response. Cell Host Microbe 12:266–275. doi:10.1016/j.chom.2012.08.005

Costa SCP, Chavez CV, Jubelin G et al (2010) Recent insight into the pathogenicity mechanisms of the emergent pathogen *Photorhabdus asymbiotica*. Microbes Infect 12:182–189. doi:10.1016/j.micinf.2009.12.003

Daborn PJ, Waterfield N, Silva CP et al (2002) A single Photorhabdus gene, makes caterpillars floppy (mcf), allows *Escherichia coli* to persist within and kill insects. Proc Natl Acad Sci U S A 99:10742–10747. doi:10.1073/pnas.102068099

Dowling AJ, Hodgson DJ (2014) An unbiased method for clustering bacterial effectors using host cellular phenotypes. Appl Environ Microbiol 80:1185–1196. doi:10.1128/AEM.03290-13

Dowling AJ, Daborn PJ, Waterfield NR et al (2004) The insecticidal toxin Makes caterpillars floppy (Mcf) promotes apoptosis in mammalian cells. Cell Microbiol 6:345–353

Dowling AJ, Waterfield NR, Hares MC et al (2007) The Mcf1 toxin induces apoptosis via the mitochondrial pathway and apoptosis is attenuated by mutation of the BH3-like domain. Cell Microbiol 9:2470–2484. doi:10.1111/j.1462-5822.2007.00974.x

Duprez L, Wirawan E, Vanden Berghe T, Vandenabeele P (2009) Major cell death pathways at a glance. Microbes Infect 11:1050–1062. doi:10.1016/j.micinf.2009.08.013

Fullner Satchell KJ (2007) MARTX, multifunctional autoprocessing repeats-in-toxin toxins. Infect Immun 75:5079–5084

Gerrard JG, Joyce SA, Clarke DJ et al (2006) Nematode symbiont for *Photorhabdus asymbiotica*. Emerg Infect Dis 12:1562–1564. doi:10.3201/eid1210.060464

Gruenheid S, Finlay BB (2003) Microbial pathogenesis and cytoskeletal function. Nature 422:775–781. doi:10.1038/nature01603

Hurst MRH, Glare TR, Jackson TA (2004) Cloning *Serratia entomophila* antifeeding genes—a putative defective prophage active against the grass grub *Costelytra zealandica*. J Bacteriol 186:5116–5128. doi:10.1128/JB.186.15.5116-5128.2004

Jank T, Bogdanović X, Wirth C et al (2013) A bacterial toxin catalyzing tyrosine glycosylation of Rho and deamidation of Gq and Gi proteins. Nat Struct Mol Biol 20:1273–1280. doi:10.1038/nsmb.2688

Lamkanfi M, Dixit VM (2010) Manipulation of host cell death pathways during microbial infections. Cell Host Microbe 8:44–54

Mulley G, Beeton ML, Wilkinson P et al (2015) From insect to man: photorhabdus sheds light on the emergence of human pathogenicity. PLoS ONE. doi:10.1371/journal.pone.0144937

Popoff MR (2014) Bacterial factors exploit eukaryotic Rho GTPase signaling cascades to promote invasion and proliferation within their host. Small GTPases 5:e28209. doi:10.4161/sgtp.28209

Roig FJ, González-Candelas F, Amaro C (2011) Domain organization and evolution of multifunctional autoprocessing repeats-in-toxin (MARTX) toxin in vibrio vulnificus. Appl Environ Microbiol 77:657–668. doi:10.1128/AEM.01806-10

Visschedyk DD, Perieteanu A, Turgeon ZJ et al (2010) Photox, a novel actin-targeting mono-ADP-ribosyltransferase from *Photorhabdus luminescens*. J Biol Chem 2:13525–13534. doi:10.1074/jbc.M109.077339

Vlisidou I, Dowling AJ, Evans IR et al (2009) Drosophila embryos as model systems for monitoring bacterial infection in real time. PLoS Pathog 5:e1000518. doi:10.1371/journal.ppat.1000518

Vlisidou I, Eleftherianos I, Dorus S et al (2010) The KdpD/KdpE two-component system of *Photorhabdus asymbiotica* promotes bacterial survival within *M. sexta* hemocytes. J Invertebr Pathol 105:352–362. doi:10.1016/j.jip.2010.09.020

Waterfield NR, Daborn PJ, Dowling AJ et al (2003) The insecticidal toxin makes caterpillars floppy 2 (Mcf2) shows similarity to HrmA, an avirulence protein from a plant pathogen. FEMS Microbiol Lett 229:265–270

Waterfield NR, Sanchez-Contreras M, Eleftherianos I et al (2008) Rapid Virulence Annotation (RVA): identification of virulence factors using a bacterial genome library and multiple invertebrate hosts. Proc Natl Acad Sci U S A 105:15967–15972. doi:10.1073/pnas.0711114105

Wood W, Jacinto A (2007) Drosophila melanogaster embryonic haemocytes: masters of multitasking. Nat Rev Mol Cell Biol 8:542–551. doi:10.1038/nrm2202

Wood W, Faria C, Jacinto A (2006) Distinct mechanisms regulate hemocyte chemotaxis during development and wound healing in *Drosophila melanogaster*. J Cell Biol 173:405–416. doi:10.1083/jcb.200508161
Yang G, Dowling AJ, Gerike U, Waterfield NR (2006) Photorhabdus virulence cassettes confer injectable insecticidal activity against the wax moth. J Bacteriol. doi:10.1128/JB.188.6.2254

Flagellar Regulation and Virulence in the Entomopathogenic Bacteria—*Xenorhabdus nematophila* and *Photorhabdus luminescens*

Alain Givaudan and Anne Lanois

Abstract There is a complex interplay between the regulation of flagellar motility and the expression of virulence factors in many bacterial pathogens. Here, we review the literature on the direct and indirect roles of flagellar motility in mediating the tripartite interaction between entomopathogenic bacteria (*Photorhabdus* and *Xenorhabdus*), their nematode hosts, and their insect targets. First, we describe the swimming and swarming motility of insect pathogenic bacteria and its impact on insect colonization. Then, we describe the coupling between the expression of flagellar and virulence genes and the dynamic of expression of the flagellar regulon during invertebrate infection. We show that the flagellar type 3 secretion system (T3SS) is also an export apparatus for virulence proteins in X. nematophila. Finally, we demonstrate that phenotypic variation, a common property of the bacterial symbionts of nematodes, also alters flagellar motility in *Photorhabdus* and *Xenorhabdus*. Finally, the so-called phenotypic heterogeneity phenomenon in the flagellar gene expression network will be also discussed. As the main molecular studies were performed in *X. nematophila,* future perspectives for the study of the interplay between flagellum and invertebrate interactions in *Photorhabdus* will be discussed.

Contents

A. Givaudan (✉) · A. Lanois
INRA, UMR Diversité, Génomes et Interactions Microorganismes-Insectes (DGIMI), Université Montpellier, CC 54, 34095 Montpellier, Cedex 05, France
e-mail: givaudan@univ-montp2.fr

A. Givaudan · A. Lanois
Université Montpellier (UM), UMR DGIMI, Montpellier, France

Current Topics in Microbiology and Immunology (2017) 402:39–52
DOI 10.1007/82_2016_53

Published Online: 26 December 2016

1 Bacterial Motility and Virulence in Pathogenic Bacteria

Many bacteria are motile through the action of large complex protein assemblages called flagella. Flagellum-mediated motility often plays an essential role in colonization by facilitating bacterial motility in mediating different bacterial interactions with plants, vertebrates, and invertebrates. In addition to motility, flagellum per se can participate in adhesion, invasion, or biofilm formation (Josenhans and Suerbaum 2002). Flagellin or depolymerised flagellum is recognized by the innate immune system in organisms as diverse as flies, plants, and mammals. Specific domains of flagellin trigger innate immune response through TLR5 and FLS2 receptors in mammals and plants, respectively (Josenhans and Suerbaum 2002). Flagellar transcriptional regulators can also contribute to virulence by regulating the expression of virulence factors and, as discussed thereafter, the flagellum is able to function as an export apparatus that mediates extracellular secretion of non-flagellar effector proteins.

A complex interplay between flagella and virulence has been demonstrated in entomopathogenic bacteria such as *Bacillus thuringiensis* (Ghelardi et al. 2002), *Xenorhabdus nematophila* (Givaudan and Lanois 2000), *Photorhabdus luminescens* (Easom and Clarke 2008), and *P. temperata* (Hurst et al. 2015) (see Table 1).

This chapter is focused on describing the interplay between bacterial flagellar-driven motility (see Sect. 2) and the tripartite interaction between bacteria (*Photorhabdus* and its sister species, *Xenorhabdus*), the nematode hosts, and the insect targets. The more detailed scientific studies on this topic have been realized in *X. nematophila.* They depicted the regulation of flagellar gene expression (in vitro and during invertebrate infection, see Sect. 3 and Sect. 5) and the flagellar T3SS (type 3 secretion system) as an export apparatus for virulence proteins (Sect. 4). Phenotypic variation is a common property of the bacterial symbionts of nematodes and here we show that this phenomenon also alters flagellar motility in both *Photorhabdus* and *Xenorhabdus* bacteria. In addition, the so-called phenotypic noise phenomenon in the flagellar gene expression network will be discussed (see Sect. 6). Future perspectives for the study of the interplay between flagellum and invertebrate interactions in *Photorhabdus* will be proposed.

2 Flagella Affect Virulence by Facilitating Motility

Flagella-driven swimming motility is a self-propulsion movement through water-filled channels within the agar (0.2–0.4% agar). Flagella can also allow one particular type of surface motility, the so-called swarming motility. Swarming has

Table 1 Impact of flagellar gene mutations on virulence against insects

Flagellar mutant strains	Phenotypes	Impact on the interaction	References
P. luminescens			
flgG *motAB*	Impaired in swarming an swimming motility (*flgG* and *motAB*)	Flagellum and motility are not required for pathogenicity or mutualism	Easom and Clarke (2008),
astR	Impaired in motility, pleiotropic mutant		Derzelle et al. (2004)
P. temperata			
flgE	Impaired in swarming and swimming motility	Slight reduced virulence	Michaels and Tisa (2011),
flgK		no impact	(Hurst et al. 2015)
X. nematophila			
fliC	Impaired in swarming and swimming motility	No impact on virulence	(Herbert and Goodrich-Blair 2007),
flhD		Attenuated virulence	(Givaudan and Lanois 2000),
fliA	All these mutants (*flhD, fliA, fliZ*) are impaired in motility and are pleiotropic	No impact on virulence	Park and Forst (2006),
fliZ		Attenuated virulence	Lanois et al. (2008)

been studied extensively in *Proteus mirabilis,* and swarmer cells are usually hyper-flagellated and move in a cell groups called rafts, and bacteria maintain close contact with other swarmer cells (Allison et al. 1992). It has been demonstrated that *P. luminescens*, *P. temperata,* and *Xenorhabdus* spp. display swimming and swarming (Derzelle et al. 2004a; Givaudan et al. 1995; Michaels and Tisa 2011). The macroscopic swarming pattern of *X. nematophila* and *P. temperata* colonies displays concentric zonation cells or "fried egg" appearance when grown on low agar concentrations (about 0.8–1%). *Xenorhabdus* and *Photorhabdus* use peritrichous flagella for both swimming and swarming motility. As expected, mutations in genes encoding flagellar structural proteins such as *flgE* (flagellar hook protein) and FlgK (flagellar hook-associated protein 1) in *P. temperata* (Hurst et al. 2015; Michaels and Tisa 2011), *flgG* (component of the distal rod) and *motAB* (flagellar motor proteins) in *P. luminescens* (Easom and Clarke 2008) and *fliC* (flagellin) in *X. nematophila* (Herbert and Goodrich-Blair 2007) were nonmotile for both swimming and swarming motility.

The assessment of the role of flagellar-driven motility per se could be achieved by the construction of mutants altering merely bacterial locomotion. To study the impact of flagellar-driven motility in the *Photorhabdus*–nematode–insect tripartite association, Easom and Clarke (2008) constructed deletion mutants of *flgG* (blocking flagella production) and *motAB* (blocking flagella rotation) in *P. luminescens* TT01. Virulence, nematode growth, and development were not impaired in non-motile mutants suggesting that flagellum and motility are not required for pathogenicity or mutualism. However, authors showed that both the ΔflgG and the ΔmotAB mutants of *P. luminescens* are out-competed by the wild-type strain illustrating that motile bacteria have a slight competitive fitness advantage during colonization of the insect larvae (Easom and Clarke 2008). Moreover, non-motile *fliC* mutant of *X. nematophila* (which lacks the flagellin subunit) has been shown to display wild-type virulence in insects (Herbert and Goodrich-Blair 2007). Thus, flagellar-driven motility per se (swarming or swimming) is not required for the virulence of EPN symbionts following injection into insect larvae but bacterial motility may be involved in more subtle advantage during bacterial colonization.

3 Flagellar Regulators Affect Virulence by Regulating Non-flagellar Virulence Factors

For a long time, it has been recognized that the flagellar regulatory proteins were specific transcription factors only for flagellar component genes, having a restricted role in swimming motility. However, in the early 2000s, flagellar transcriptional regulators were shown to control the expression of non-flagellar genes in *Escherichia coli* and *Xenorhabdus* (Givaudan and Lanois 2000; Pruss et al. 2003). Actually, more than 50 genes are involved in the biogenesis and function of a flagellum in *E. coli* or *Salmonella* Typhimurium. These genes are transcriptionally regulated as a cascade and are coordinated with the flagellar hierarchy. At the top of the hierarchy is the class I operon, *flhDC*, whose products are required for the expression of all other flagellar genes (see Fig. 1). The *E. coli* FlhD and FlhC proteins act as an activator for class II operons including most of the structural genes for the flagellar hook–basal body complexes plus the alternative sigma factor *fliA*. The product of the *fliA* gene, σ28, directs the transcription of class III genes that encode the filament protein, hook-associated proteins, motor proteins, and various chemotaxis proteins. The central channel is believed to work as a passage not only for flagellar component proteins, but also for flagellar regulatory protein FlgM, an anti-sigma factor. Accumulation of FlgM in the cell by preventing its export blocks the transcription. Two other genes within the flagellar regulon, *fliT* and *fliZ*, have been shown to regulate class II gene transcription (Kalir et al. 2001).

First in *E. coli*, it has been demonstrated that FlhD, the flagellar master regulator, is involved in processes other than flagellar expression that occurs when cells enter the stationary phase (Pruss and Matsumura 1996). In addition, by using *flhD* mutants and phenotypic characterization, Givskov et al. (1995) and Young et al. (1999) have shown that the *flhDC* operon controls phospholipase expression *in Serratia*

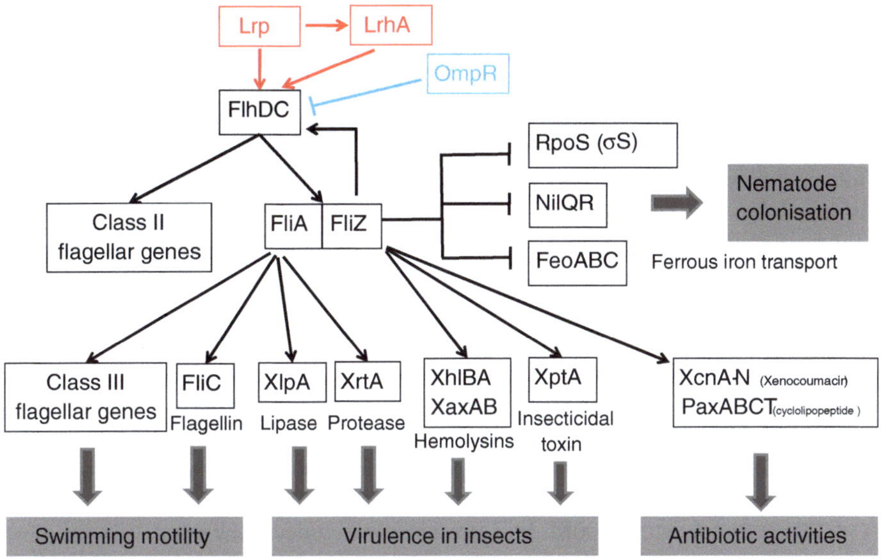

Fig. 1 Model of flagellar regulation cascade in *X. nematophila*. In this regulatory hierarchy model, arrows and crossbars indicate positive and negative regulations, respectively, that are either direct or indirect. Direct binding has been demonstrated for FliZ at *flhDC* and hemolysin (*xaxAB* and *xhlBA*) promoters (Lanois et al. 2008). All other regulatory relationships are inferred from genetic or bioinformatic analyses. For simplicity, some known regulatory connections are not shown. In this model, the flagellar master operon *flhDC* is positively regulated by Lrp and LrhA (*red boxes* and *arrows*) (Richards et al. 2008) and negatively regulated by OmpR (*blue boxes* and *arrows*) (Park and Forst 2006). Class II–III flagellar genes (strictly FlhD-dependent) and exoprotein genes (lipase, protease, and hemolysins) are coregulated through *fliAZ* operon products (Givaudan and Lanois 2000; Park and Forst 2006). FliZ has been shown to directly control the *flhDC* operon through a positive feedback (Lanois et al. 2008). In addition, FliZ directly or indirectly upregulates the expression of genes encoding insecticidal toxins and antimicrobial compounds and downregulates the expression of two genes encoding the transcriptional regulators RpoS and NilR, and a structural gene, *feoABC* (Jubelin et al. 2013)

liquefaciens and *Yersinia enterocolitica,* respectively. We previously showed that FlhDC positively regulates swarming behavior, hemolysin, and lipase production in the insect pathogenic bacteria, *X. nematophila* (Givaudan and Lanois 2000). In addition, we demonstrated that FlhDC regulatory proteins of *Xenorhabdus* are involved in host interactions being essential for full virulence after injection into insects. In *X. nematophila*, the transcription of *flhDC* is also negatively influenced by the two-component regulator EnvZ/OmpR (Park and Forst 2006) and is positively regulated by LrhA (Richards et al. 2008) and the global regulator Lrp (Cowles et al. 2007) (see Fig. 1). Consequently, mutations in the genes encoding these regulators have also been shown to affect swimming and swarming motilities, lipase and protease production, and hemolysis.

In this manner, the main regulatory pathway controlling these coregulated phenotypes was identified, and the precise hierarchy in the flagellar cascade was documented. First, FliA, the sigma 28 factor, co-ordinates the expression of two non-flagellar genes, *xlpA* and *xrtA*, encoding, respectively, a lipase and a protease and the flagellar motility (Park and Forst 2006). Next, another flagellar regulator, FliZ, is encoded by the second gene in the *fliAZ* operon. Surprisingly, the motility displayed by the *Xenorhabdus fliZ* mutant was strongly altered relative to the wild-type strain, whereas the *fliZ* mutants of *E. coli* and *S.* Typhimurium are fully motile. However, unlike the polar *fliAZ* mutation, *fliZ* deletion did not fully abolish the swimming capacity of *Xenorhabdus* (Jubelin et al. 2013; Lanois et al. 2008). A detailed molecular study was undertaken to identify the gene network controlled by the *X. nematophila* flagellar regulator FliZ. A global RNA-Seq analysis demonstrated that FliZ controls 278 coding sequences (Jubelin et al. 2013). FliZ has a positive impact on the expression of all the genes belonging to the flagellar cascade. This effect is due to the FliZ positive feedback loop on *flhDC* expression and the levels of FlhDC are critical for efficient motility in *Xenorhabdus* (Jubelin et al. 2013; Lanois et al. 2008). FliZ up- or downregulated the expression of genes encoding many non-flagellar proteins potentially involved in key steps of the *Xenorhabdus* life cycle. FliZ was found to upregulate hemolysin gene expression by binding directly to the *xaxAB* and *xhlAB* promoter regions (Lanois et al. 2008). The two directly FliZ-dependent genes, termed *xaxAB* and *xhlBA*, encode hemolysins from two different families. XaxAB is the prototype of a new extensive family of hemolysins with apoptotic and pore-forming activities in mammalian and invertebrate cells (Vigneux et al. 2007). The second FliZ-dependent hemolysin, XhlA, belongs to a family characterized by a two-partner secretion system (TPSS). This cell surface-associated hemolysin has been shown to be required for the full virulence of *X. nematophila* in insect larvae (Cowles and Goodrich-Blair 2005). Moreover, through positive feedback on *fliAZ* expression, FliZ also modulates the expression of the FliA-dependent gene *prtA* (also called *xrtA*), which encodes a protease (Park and Forst 2006). Another FliZ-dependent gene is *xptA*, a gene encoding a high molecular weight toxin complex (Tc) protein with insecticidal effects found in *Xenorhabdus* and *Photorhabdus* (Waterfield et al. 2001). Therefore, it is likely that downregulation of the expression of genes encoding the hemolysins and Tc toxins observed in the *fliZ* mutant explains the delayed virulence pattern observed with the *fliZ* mutant (Jubelin et al. 2013). In conclusion, FliZ has been shown to mediate the coordinate regulation of flagellum synthesis and virulence in the insect pathogen *X. nematophila*.

In *P. luminescens*, no mutants in flagellar regulators have been described. However, mutations in global transcriptional regulators such as the two-component system, PhoPQ (Derzelle et al. 2004b), and AstR affect the ability of *Photorhabdus* cells to swim and swarm on agar surfaces. Further, a transcriptional analysis in *P. luminescens* indicated that the negative regulation by AstR occurs at the *flhDC* transcription level and explains the early onset of swarming observed in the *astR* mutant. However, this *astR* mutation has no apparent impact on virulence in insects (Derzelle et al. 2004a).

4 The Flagella Apparatus as Virulence Factor Secretion System

Sequence similarities exist between components of bacterial type III secretion systems (T3SS) and those of the flagellar assembly machinery (Blocker et al. 2003). T3SSs are essential determinants of the interaction of many Gram-negative bacteria with animal or plant cells and facilitate the translocation of bacterial proteins into eukaryotic host cells. Similarly, the flagellum has also been shown to function as an export apparatus that mediates extracellular secretion of non-flagellar virulence-associated effector proteins. In contrast to *P. luminescens* (Brugirard-Ricaud et al. 2005), it is noteworthy that *X. nematophila* does not encode a dedicated (i.e., non-flagellar) type III secretion system (Brugirard-Ricaud et al. 2004). As described above, *xlpA* (lipase) and *xrtA* (protease) are both regulated by FliA-dependent promoters in *X. nematophila* in a similar way to class III flagellar genes. Therefore, the question remained on the secretory pathways responsible for enzyme export. One way to solve this question is the use of mutants in export apparatus components, such as FlhA and FlhB. Indeed loss of lipase, but not protease, activity was reported in the *flhA* strain of *X. nematophila* suggesting that XlpA is secreted via the flagellar export apparatus while XrtA is secreted by a dedicated ABC transport system (Park and Forst 2006). Moreover, the FliZ-dependent XaxAB hemolysin activity was not abolished in the *flhA* strain indicates that this hemolysin is secreted by an unknown pathway other than the flagellar export system. It was also recently proposed that the type III export apparatus of the flagellar system transports virulence-associated phospholipase A in *Y. enterocolitica* (Young et al. 1999) and *Serratia liquefaciens* (Givskov et al. 1995). Finally, the expression and secretion of lipase XlpA in *Xenorhabdus* is dependent on a functional flagellar system.

The bacteriocin xenocin complex is also exported through the flagellar type III system in *X. nematophila*. Indeed, the xenocin operon of *X. nematophila* consists of *xciA* and *ximB* genes encoding a xenocin and immunity protein able to kill competing microbes in the insect. The expression of xenocin operon is not dependent on flagellar regulators but is dependent on SOS induction. Using an *flhA* strain, authors suggested the involvement of the flagellar type III secretory system in xenocin export (Singh et al. 2013). These data demonstrate a surprising uncoupling between expression and secretion of this non-flagellar protein.

5 In Situ Flagellar Gene Expression During Invertebrate Infection

The *X. nematophila* life cycle has been divided into three phases: (1) insect infection, (2) bacterial and nematode reproduction in the insect cadaver, and (3) nematode transmission of *X. nematophila* to insect hosts (Richards and Goodrich-Blair 2009). Gene inactivation studies have identified regulatory proteins involved in adaptation to the shifting between these different hosts. Based on such data, Richard and Goodrich-Blair (2009) proposed a regulatory model that highlights the temporal changes in symbiotic and virulence factor expression that may allow *X. nematophila* to cope with the shifts in host environments inherent to the successful colonization of the two invertebrate hosts. However, few studies have investigated gene expression during transitions between *X. nematophila* life stages. Jubelin et al. (2011) investigated the dynamics of the FliA-dependent flagellin gene *fliC* and FliZ-dependent hemolysin genes (*xaxAB* and *xhlBA*) during insect infection and nematode association by carrying out real-time expression analysis using an unstable GFP monitoring system. Regardless of the route of infection (bacterial injection or infestation by EPN complexes), expression of flagellar genes and flagellar regulated-hemolysin genes was high only in dead insects (see Fig. 2A) and these genes were not expressed in infective juvenile stages of nematodes (see Fig. 2B). It has been proposed that activation of the flagellar cascade in dead insects may allow bacteria to move through cadaver tissues toward nutrient-rich niches by chemotaxis, facilitating their growth. In addition, other activities (lipases, proteases, and hemolysins) controlled by the flagellar regulon may also contribute to the degradation of insect tissues, releasing nutrients required for nematode feeding. Also, the expression of FliZ-dependent hemolysin genes coincided with the increase in iron availability detected at the time of insect death, suggesting that iron availability is a signal governing the adaptation of *X. nematophila* to changes in host environments. This study demonstrated for the first time that the flagellar regulon in *X. nematophila* is transiently expressed for the period of the shift

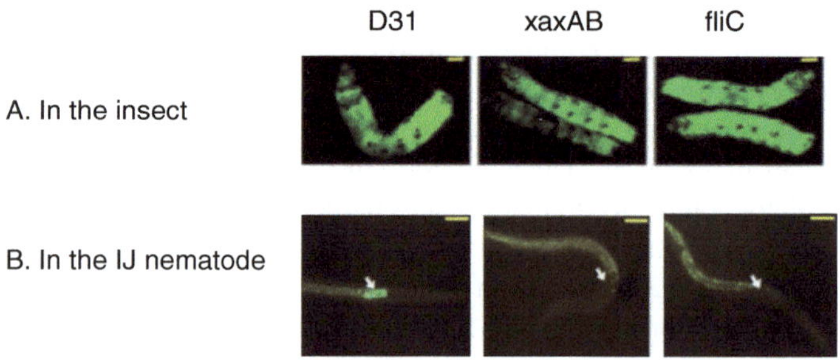

Fig. 2 Expression of hemolysin and flagellin genes within infected insects and infective juvenile (IJ) nematodes. **A** Temporal and spatial expression of hemolysin and flagellin genes during the late stage of insect infection. For each experiment, at time zero, we injected 10^4 bacteria (F1 strains carrying PD31–*gfp*[AAV], P*xaxAB*–*gfp*[AAV] and P*fliC*–*gfp*[AAV] constructs) into the hemocoel of the insect larvae. Whole larvae were observed with a fluorescence macroscope, with an exposure time set to 600 ms for photography (bars, 2.5 mm). **B** The flagellin and hemolysin genes are not expressed in *X. nematophila* bacteria associated with IJ nematodes. Axenic nematodes of *S. carpocapsae* were associated with F1 strains carrying PD31–*gfp*[AAV], P*xaxAB*–*gfp*[AAV], and P*fliC*–*gfp*[AAV] constructs. Larvae of *S. littoralis* were infested with the various bacterium–nematode complexes and IJs were observed by fluorescence microscopy shortly after their emergence from insect cadavers. Arrows indicate the bacterium-containing vesicle of the IJs. Bars represent 30 mm. Adapted from Jubelin et al. (2011)

between the insect infection and nematode reproduction phases. Interestingly, this paper also revealed that the expression of the flagellar and hemolysin genes in *Xenorhabdus* was heterogeneous at the individual cell level (see Sect. 6).

6 Phenotypic Variation and Bimodal Expression of Flagellar and Virulence Genes

All *Xenorhabdus* species and both *P. luminescens* and *P. temperata* have been reported to undergo "phenotypic variation" characterized by a switching between two cell types known as "primary" and "secondary" forms or variants. Although the phenotypic differences between primary and secondary forms can vary depending on strain and species (Hurlbert et al. 1989; Smigielski et al. 1994), typically the primary, but not secondary, cells are more highly pigmented, agglutinate red blood cells, produce fimbriae, hemolysins, proteases, antimicrobials, and crystalline inclusion bodies, and are impaired in bioluminescence for *Photorhabdus* (Gerritsen et al. 1995; Givaudan et al. 1995; Hurlbert et al. 1989; Smigielski et al. 1994; Volgyi et al. 2000). The swimming motility of variants in *P. temperata* was also investigated (Michaels and Tisa 2011) but both the primary and the secondary variants were able to swim in liquid or semisolid media under appropriate conditions. However, variation in the oxygen level greatly influenced the behavior of the secondary form and not motility of primary form. Also, it is demonstrated that the LysR-type transcriptional regulator HexA from *P. temperata* is able to repress several primary-specific phenotypes, i.e., exoenzyme production, pigmentation, and antibiotic activity but no effect was observed on motility (Joyce and Clarke 2003). In *P. luminescens,* a study showed that mutation in *astRS* genes, encoding a two-component signal transduction system, induces an earlier transition to the secondary phenotype than the wild type (Derzelle et al. 2004a). This mutation also affected motility in *P. luminescens.*

In *Xenorhabdus,* (Givaudan et al. 1995) revealed that most primary variants exhibit swimming and swarming behavior, whereas secondary variants in *X. nematophila* and *X. bovienii* are non-motile. Exhaustive molecular studies in *X. nematophila* showed that the secondary form was unable to synthesize flagellar filaments and that the flagellin-encoding gene, *fliC*, and the hook-associated protein 2 gene, *fliD*, were switched off at the transcriptional level in the variants (Givaudan et al. 1996). These results suggested that the expression of a gene earlier in the transcriptional hierarchy of the flagellar regulon was impaired. However, it showed that *flhDC* gene structure and expression of the master regulator FlhDC are not altered in variants (Givaudan and Lanois 2000), suggesting that this locus is not responsible for flagellar variation phenomenon in *X. nematophila*.

Another phenomenon observed in the flagellar regulon of *X. nematophila* is the so-called "phenotypic heterogeneity". This phenomenon describes usually non-genetic variations that are observed between individual cells in an isogenic population (Veening et al. 2008). The terms "bistability" or "bimodality" describe the situation in which the population has bifurcated into coexisting cell types one representing the "OFF" state, and the other representing the "ON" state. Using single-cell analysis, (Jubelin et al. 2013) showed a bimodal expression of FliZ-dependent genes (class II–III flagellar genes and hemolytic genes) during exponential growth of the bacterial population. Actually, this bimodality generated a mixed population of subsets of cells in which FliZ-dependent genes are expressed at high- or low-level cells either expressing ("ON state") or not expressing ("OFF state") FliZ-dependent genes (see Fig. 3). It is known that in the presence of positive regulatory feedback, a graded expression can be converted into a binary response, in which cells express a certain gene at high or low levels (Veening et al. 2008). In addition, studies of a bacterial population exposed to a graded series of FliZ concentrations showed that FliZ controls the rate of transition between the "OFF" and "ON" states in individuals (see Fig. 3). FliZ thus plays a key role in cell fate decisions, by transiently creating individuals with different potentials for motility and host interaction genes encoding non-flagellar proteins potentially involved in key steps of the *Xenorhabdus* life cycle (see Fig. 1). It is likely that the heterogeneity already observed in the course of insect infection (see Sect. 5) is due to this bimodality phenomenon.

The relationship between the FliZ-mediated bimodality and the phenotypic variation in *X. nematophila* is still therefore an open question. In a preliminary

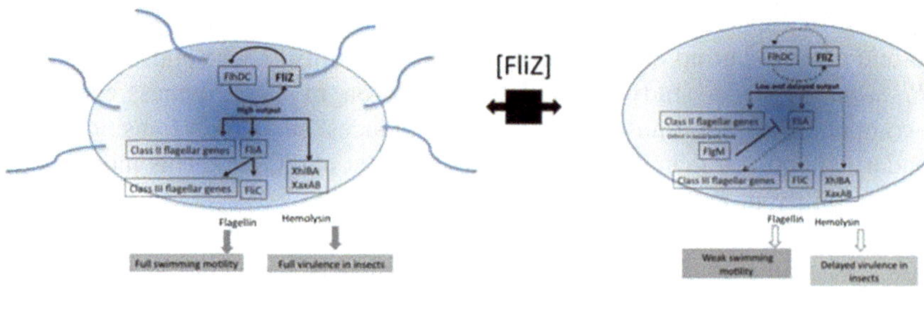

Fig. 3 Model depicting the regulatory network underlying the regulation of flagella and the coregulation of insect virulence. The diagram shows a mixed bacterial population coexisting during the exponential growth phase of *X. nematophila*. On the *left*, a motile bacterium is shown. When the stochastic expression of the circuit generates large amounts of FliZ, this molecule exerts positive feedback on *flhDC* expression, upregulating the flagellar cascade and FliZ-dependent hemolysin genes. Consequently, the noisy expression of class II flagellar genes is reduced and cells fully express classes II and III flagellar genes, resulting in the motility phenotype. On the right, a non-motile bacterium is shown. Authors suggest that the lower level of FlhD-FliZ output delays and desynchronizes class II genes expression, probably impairing completion of the basal body–hook structure. The FlgM protein, an anti-sigma-28 factor, binds the FliA sigma-28 factor directly in *E. coli*, preventing class III promoter transcription until after hook–basal body completion (Kutsukake 1994). Thus, the accumulation of FlgM in cells probably blocks the transcription of class III genes, including that encoding flagellin, resulting in a non-motile state. The threshold of FliZ controls the rate of transition between the "OFF" and "ON" states in individuals in *Xenorhabdus*. adapted from Jubelin et al. (2013)

experiment, we investigated the form status (primary or secondary forms) in ON and OFF populations after cell sorting, but no variant forms were identified (our unpublished data) even in the OFF population in which transient expression patterns were similar to that observed in the secondary form.

7 Conclusions and Future Work on *P. luminescens*

Flagellar motility plays an important role in the life cycle of the entopathogenic bacteria. However for *Photorhabdus* and *Xenorhabdus*, it is clear that the bacterial locomotion per se has little impact on the interaction with nematodes or insects. However, the coupling of motility with other bacterial virulence factors appears to be mediated in part by master regulatory proteins in motility systems (FlhDC, FliA, and FliZ) of *X. nematophila*. It is therefore clear that virulence and motility are coordinately regulated to best support the lifestyle of the pathogenic bacteria in the course of insect infection. Further research will improve our understanding of how virulence and motility are coordinately regulated in *Photorhabdus*. One additional aspect which has to be investigated further is the role of flagella in innate immunity of insects. While flagellin triggers humoral immunity in fly Samakovlis et al. (1992), no flagellin TLR-like ligands have yet been identified in other insects. Finally, the switch-off of the expression of the flagellar regulon during the colonization of nematode infective juvenile larva suggests that the flagellar regulon is not involved in mutualism in *Xenorhabdus*. However, this question remains open for the complex association between *Photorhabdus* and *Heterorhabditis*.

References

Allison C, Lai HC, Hughes C (1992) Co-ordinate expression of virulence genes during swarm-cell differentiation and population migration of *Proteus mirabilis*. Mol Microbiol 6:1583–1591

Blocker A, Komoriya K, Aizawa S (2003) Type III secretion systems and bacterial flagella: insights into their function from structural similarities. Proc Natl Acad Sci USA 100:3027–3030. doi:10.1073/pnas.0535335100

Brugirard-Ricaud K, Givaudan A, Parkhill J, Boemare N, Kunst F, Zumbihl R, Duchaud E (2004) Variation in the effectors of the type III secretion system among *Photorhabdus* species as revealed by genomic analysis. J Bacteriol 186:4376–4381. doi:10.1128/JB.186.13.4376-4381.2004

Brugirard-Ricaud K, Duchaud E, Givaudan A, Girard PA, Kunst F, Boemare N, Brehelin M, Zumbihl R (2005) Site-specific antiphagocytic function of the *Photorhabdus luminescens* type III secretion system during insect colonization. Cell Microbiol 7:363–371. doi:10.1111/j.1462-5822.2004.00466.x

Cowles KN, Goodrich-Blair H (2005) Expression and activity of a *Xenorhabdus nematophila* haemolysin required for full virulence towards *Manduca sexta* insects. Cell Microbiol 7:209–219

Cowles KN, Cowles CE, Richards GR, Martens EC, Goodrich-Blair H (2007) The global regulator Lrp contributes to mutualism, pathogenesis and phenotypic variation in the bacterium *Xenorhabdus nematophila*. Cell Microbiol 9:1311–1323

Derzelle S, Ngo S, Turlin E, Duchaud E, Namane A, Kunst F, Danchin A, Bertin P, Charles JF (2004a) AstR-AstS, a new two-component signal transduction system, mediates swarming, adaptation to stationary phase and phenotypic variation in *Photorhabdus luminescens*. Microbiology 150:897–910. doi:10.1099/mic.0.26563-0

Derzelle S, Turlin E, Duchaud E, Pages S, Kunst F, Givaudan A, Danchin A (2004b) The PhoP-PhoQ two-component regulatory system of *Photorhabdus luminescens* is essential for virulence in insects. J Bacteriol 186:1270–1279

Easom CA, Clarke DJ (2008) Motility is required for the competitive fitness of entomopathogenic *Photorhabdus luminescens* during insect infection. BMC Microbiol 8:168. doi:10.1186/1471-2180-8-168

Gerritsen L, van der Wolf JM, van Vuurde J, Ehlers R, Krasomil-Osterfel KC, Smits PH (1995) Polyclonal Antisera to distinguish strains and form variants of *Photorhabdus (Xenorhabdus) luminescens*. Appl Environ Microbiol 61:284–289

Ghelardi E, Celandroni F, Salvetti S, Beecher DJ, Gominet M, Lereclus D, Wong AC, Senesi S (2002) Requirement of flhA for swarming differentiation, flagellin export, and secretion of virulence-associated proteins in *Bacillus thuringiensis*. J Bacteriol 184:6424–6433

Givaudan A, Lanois A (2000) *flhDC*, the flagellar master operon of *Xenorhabdus nematophilus*: requirement for motility, lipolysis, extracellular hemolysis, and full virulence in insects. J Bacteriol 182:107–115

Givaudan A, Baghdiguian S, Lanois A, Boemare N (1995) Swarming and swimming changes concomitant with phase variation in *Xenorhabdus nematophilus*. Appl Environ Microbiol 61:1408–1413

Givaudan A, Lanois A, Boemare N (1996) Cloning and nucleotide sequence of a flagellin encoding genetic locus from *Xenorhabdus nematophilus*: phase variation leads to differential transcription of two flagellar genes (fliCD). Gene 183:243–253

Givskov M, Eberl L, Christiansen G, Benedik MJ, Molin S (1995) Induction of phospholipase- and flagellar synthesis in *Serratia liquefaciens* is controlled by expression of the flagellar master operon *flhD*. Mol Microbiol 15:445–454

Herbert EE, Goodrich-Blair H (2007) Friend and foe: the two faces of Xenorhabdus nematophila. Nat Rev Microbiol 5:634–646

Hurlbert RE, Xu J, Small CL (1989) Colonial and cellular polymorphism in *Xenorhabdus luminescens*. Appl Environ Microbiol 55:1136–1143

Hurst S, Rowedder H, Michaels B, Bullock H, Jackobeck R, Abebe-Akele F, Durakovic U, Gately J, Janicki E, Tisa LS (2015) Elucidation of the *Photorhabdus temperata* Genome and generation of a transposon mutant library to identify motility mutants altered in Pathogenesis. J Bacteriol 197:2201–2216. doi:10.1128/JB.00197-15

Josenhans C, Suerbaum S (2002) The role of motility as a virulence factor in bacteria. Int J Med Microbiol 291:605–614

Joyce SA, Clarke DJ (2003) A hexA homologue from *Photorhabdus* regulates pathogenicity, symbiosis and phenotypic variation. Mol Microbiol 47:1445–1457

Jubelin G, Pages S, Lanois A, Boyer MH, Gaudriault S, Ferdy JB, Givaudan A (2011) Studies of the dynamic expression of the *Xenorhabdus* FliAZ regulon reveal atypical iron-dependent regulation of the flagellin and haemolysin genes during insect infection. Environ Microbiol 13:1271–1284. doi:10.1111/j.1462-2920.2011.02427.x

Jubelin G, Lanois A, Severac D, Rialle S, Longin C, Gaudriault S, Givaudan A (2013) FliZ is a global regulatory protein affecting the expression of flagellar and virulence genes in individual *Xenorhabdus nematophila* bacterial cells. PLoS Genet 9:e1003915. doi:10.1371/journal.pgen.1003915

Kalir S, McClure J, Pabbaraju K, Southward C, Ronen M, Leibler S, Surette MG, Alon U (2001) Ordering genes in a flagella pathway by analysis of expression kinetics from living bacteria. Science 292:2080–2083. doi:10.1126/science.1058758

Kutsukake K (1994) Excretion of the anti-sigma factor through a flagellar substructure couples flagellar gene expression with flagellar assembly in *Salmonella typhimurium*. Mol Gen Genet 243:605–612

Lanois A, Jubelin G, Givaudan A (2008) FliZ, a flagellar regulator, is at the crossroads between motility, haemolysin expression and virulence in the insect pathogenic bacterium *Xenorhabdus*. Mol Microbiol 68:516–533

Michaels B, Tisa LS (2011) Swarming motility by *Photorhabdus temperata* is influenced by environmental conditions and uses the same flagella as that used in swimming motility. Can J Microbiol 57:196–203. doi:10.1139/W10-119

Park D, Forst S (2006) Co-regulation of motility, exoenzyme and antibiotic production by the EnvZ-OmpR-FlhDC-FliA pathway in *Xenorhabdus nematophila*. Mol Microbiol 61:1397–1412

Pruss BM, Matsumura P (1996) A regulator of the flagellar regulon of *Escherichia coli, flhD*, also affects cell division. J Bacteriol 178:668–674

Pruss BM, Campbell JW, Van Dyk TK, Zhu C, Kogan Y, Matsumura P (2003) FlhD/FlhC is a regulator of anaerobic respiration and the Entner-Doudoroff pathway through induction of the methyl-accepting chemotaxis protein Aer. J Bacteriol 185:534–543

Richards GR, Goodrich-Blair H (2009) Masters of conquest and pillage: *Xenorhabdus nematophila* global regulators control transitions from virulence to nutrient acquisition. Cell Microbiol 11:1025–1033. doi:10.1111/j.1462-5822.2009.01322.x

Richards GR, Herbert EE, Park Y, Goodrich-Blair H (2008) *Xenorhabdus nematophila lrhA* is necessary for motility, lipase activity, toxin expression, and virulence in *Manduca sexta* insects. J Bacteriol 190:4870–4879. doi:10.1128/JB.00358-08

Samakovlis C, Asling B, Boman HG, Gateff E, Hultmark D (1992) In vitro induction of cecropin genes–an immune response in a Drosophila blood cell line. Biochem Biophys Res Commun 188:1169–1175

Singh P, Park D, Forst S, Banerjee N (2013) Xenocin export by the flagellar type III pathway in *Xenorhabdus nematophila*. J Bacteriol 195:1400–1410. doi:10.1128/JB.01532-12

Smigielski AJ, Akhurst RJ, Boemare NE (1994) Phase Variation in Xenorhabdus nematophilus and Photorhabdus luminescens: Differences in Respiratory Activity and Membrane Energization. Appl Environ Microbiol 60:120–125

Veening JW, Smits WK, Kuipers OP (2008) Bistability, epigenetics, and bet-hedging in bacteria. Annu Rev Microbiol 62:193–210. doi:10.1146/annurev.micro.62.081307.163002

Vigneux F, Zumbihl R, Jubelin G, Ribeiro C, Poncet J, Baghdiguian S, Givaudan A, Brehelin M (2007) The *xaxAB* genes encoding a new apoptotic toxin from the insect pathogen *Xenorhabdus nematophila* are present in plant and human pathogens. J Biol Chem 282:9571–9580

Volgyi A, Fodor A, Forst S (2000) Inactivation of a novel gene produces a phenotypic variant cell and affects the symbiotic behavior of *Xenorhabdus nematophilus*. Appl Environ Microbiol 66:1622–1628

Waterfield NR, Bowen DJ, Fetherston JD, Perry RD, ffrench-Constant RH (2001) The *tc* genes of *Photorhabdus*: a growing family. Trends Microbiol 9:185–191

Young GM, Schmiel DH, Miller VL (1999) A new pathway for the secretion of virulence factors by bacteria: the flagellar export apparatus functions as a protein-secretion system. Proc Natl Acad Sci U S A 96:6456–6461

Part II
Secondary Metabolism and its Regulation

Natural Products from *Photorhabdus* and Other Entomopathogenic Bacteria

Kenan A.J. Bozhüyük, Qiuqin Zhou, Yvonne Engel, Antje Heinrich, Alexander Pérez and Helge B. Bode

Abstract Although the first natural products (NP) from *Photorhabdus* and *Xenorhabdus* bacteria have been known now for almost 30 years, a huge variety of new compounds have been identified in the last 5–10 years, mainly due to the application of modern mass spectrometry. Additionally, application of molecular methods that allow the activation of NP production in several different strains as well as efficient heterologous expression methods have led to the production and validation of many new compounds. In this chapter we discuss the benefit of using *Photorhabdus* as a model system for microbial chemical ecology. We also examine non-ribosomal peptide synthetases as the most important pathway for NP production. Finally, we discuss the origin and function of all currently known NPs and the development of the molecular and chemical tools used to identify these NPs faster.

Contents

K.A.J. Bozhüyük · Q. Zhou · Y. Engel · A. Heinrich · A. Pérez · H.B. Bode (✉)
Merck Endowed Chair for Molecular Biotechnology, Department of Biosciences and Buchmann Institute for Molecular Life Sciences (BMLS), Goethe University Frankfurt, Max-von-Laue-Str. 9, 60438 Frankfurt am Main, Germany
e-mail: h.bode@bio.uni-frankfurt.de

Current Topics in Microbiology and Immunology (2017) 402:55–80
DOI 10.1007/82_2016_24

Published Online: 28 December 2016

1 Introduction

Low molecular weight natural products (NPs) from plants and other biological sources have been used by humans since ancient times. Today's modern life without such microbial natural products would be nearly impossible. For example, 30–60 % of all drugs currently in clinical use as antibiotics (e.g. vancomycin, daptomycin, penicillin, erythromycin), anti-cancer agents (e.g. doxycycline) or immune suppressant compounds (e.g. cyclosporine, rapamycin) are natural products or are directly derived from them (Newman and Cragg 2012). Despite our success in identifying these compounds and their subsequent medical utility it is still not clear why these natural products are actually produced by these microorganisms and therefore what role(s) they play in their biology. Thus, whilst it is easy to infer a role for antibiotically active NPs in growth competition and (or) defence in the natural environment. For other natural products, such as those used as cholesterol lowering agents, their biological role(s) in nature are far less clear. In this respect it is interesting to note that even 'antibiotics' induce changes in the mRNA and protein levels of other bacteria when used in sub-inhibitory concentrations (Davies and Ryan 2012; Davies et al. 2006). Therefore compounds used by us as antibiotics might in fact assume a 'signalling' function at the low concentrations found in the natural environments. Only the subsequent 'misuse' of these signalling compounds in non-natural high doses then might result in a clinically relevant antibiotic activity.

Originally, NPs from microorganisms and plants were designated as 'secondary' metabolites not essential for growth in contrast to the 'primary' metabolites considered essential for growth. The limits of this primary–secondary classification were always the specific laboratory conditions used for such analysis. For example, typically a pure culture of a single bacterium was studied without any competitor in the laboratory. However, subsequent studies of microbial ecology have revealed that typical secondary metabolites may indeed have an 'essential' role but only under specific conditions of growth and or competition (e.g. iron siderophores or quorum sensing molecules). Thus, it might be more reasonable to call such compounds 'specialized' metabolites for specific conditions we often either do not recognize or fail to classify properly (Sharon et al. 2014).

From the recent genome sequencing projects of various microorganisms it has become obvious that almost bacteria, including anaerobes, that have previously been thought to not encode any biosynthesis gene cluster (BGC) for NP production do indeed carry such BGCs (Bode and Müller 2005). In fact between 6 and 8 % of the genome can be dedicated to NP biosynthesis with up to 25 BGCs found in some *Streptomyces* strains (Ikeda et al. 2003; Omura et al. 2001). Despite the extent of the genome encoding NPs, most of them are not expressed under the conditions analysed and are therefore often called 'silent'. This point will be addressed in detail in the chapter by David Clarke also in this volume. Because of the large amount of energy and resources used to maintain NP production, with several microbes producing multiple compounds in parallel (Bode and Müller 2005), there is increasing evidence that NP production is regulated 'heterogeneously', as also found for the production of some quorum sensing compounds. In such a scenario, typical for a 'division of labour' (Ackermann 2015), only a fraction of cells actually produce a specific NP thereby reducing the workload for each single cell but also allowing the whole population to produce all NPs required at a specific time. Although, such research has just started it is evident from the sheer physical size of some NP BGCs that such a strategy would be extremely beneficial.

Probably the biggest current challenge in natural product research is the identification of the biological function of the specific NP produced. Despite recent rapid progress in microbial ecology, such as the development of single colony growth chambers and direct microbial detection (Ling et al. 2015), we are still limited by the complexity of natural communities and our ability to culture only a tiny fraction of this complex mixture of micro-organisms, in combination with a limited number of molecular tools to examine those organisms which we can indeed culture. Therefore, the real chemical ecology of NPs from typical soil microbes, like *Streptomyces*, is still a very complex black box leading people to turn to more simple systems to address basic questions about the biological role of NPs. In this respect, micro-organisms living in symbiosis might carry an advantage for such studies when all the interacting partners can be isolated and studied in the laboratory. Although such laboratory systems might not be fully 'natural' since a number of natural competitors might be absent, they do allow us to address at least one naturally relevant interaction in detail and possibly from both sides in cases where the symbiotic host is also accessible to analysis and manipulation. The ultimate goal would be to address such complex symbioses, like the interaction of humans with their microbiome, identifying the influence of individual NPs for the shaping of the microbiome itself and subsequent human developmental biology and even behaviour (Dorrestein et al. 2014). Since the human microbiome is still a major challenge, with a complexity similar to that seen in soil communities, people have started to look at lower eukaryote–microbe interactions as models (Cantley and Clardy 2015). In this respect, insects and nematodes are very promising model systems in which to tackle question of chemical ecology since they can both be cultivated easily in the laboratory and are also amenable to molecular methods.

In this chapter I will therefore focus on NPs from *Photorhabdus* bacteria which has been established as a model insect pathogen, with a few excursions to its sister taxon *Xenorhabdus*. I note that the main biosynthesis and regulation pathways for the production of *Photorhabdus* NPs will be introduced by David Clarke elsewhere in this volume.

2 Non-ribosomal Peptide Synthetases (NRPS) and Related Biosynthesis Pathways

Based on their biosynthetic pathway, peptide NPs can be divided into two classes (Grünewald and Marahiel 2006). Bacteriocines for instance are ribosomally synthesized and post-translationally modified peptides (RiPPS) with antibacterial activity often produced by bacteria to inhibit the growth of similar or closely related bacterial strains (Arnison et al. 2012). Further important subclass are non-ribosomally produced peptides, like the antibiotic daptomycin, the anti-cancer drug bleomycin and the immunosuppressant cyclosporin (Felnagle et al. 2008). The most striking difference to ribosomally synthesized peptides is their method of assembly. Non-ribosomal peptides (NRPs) are produced via large multi-enzyme complexes (mega-enzymes) using thiotemplate mechanisms like NRP synthetases (NRPS) and NRPS/PKS hybrids. They use peptide bond formation (NRPS) or a combination of peptide bond formation and Claisen-type condensation reactions (NRPS/polyketide synthase [PKS] hybrids) to build larger molecules from different building blocks, not limited to the 20 proteinogenic amino acids (Caboche et al. 2010). Further, these NRP exhibit unique structural elements, like *D*-, *N*- or *C*-methylated, glycosylated or phosphorylated amino acids, heterocycles, and/or N-terminal attached fatty acids.

A common feature of these molecules is their often constrained structure (Sieber and Marahiel 2005). For example, they exhibit macrocyclization (macrolactam or depsipeptide with one ester bond), heterocyclization or C–C or C–O cross-linking between different amino acids, which enhances the rigidity of their ring skeleton (Grünewald and Marahiel 2006). This rigidity in turn ensures bioactivity by a precise orientation required for interaction with a dedicated molecular target. Although peptides synthesized by NRPS are highly diverse in structure, most of them share a common mode of synthesis, denoted as multiple carrier thiotemplate mechanism. Further, NRPS harbour a strict 'modular' architecture. A 'module' is defined as the catalytic unit responsible for the incorporation of one specific building block (e.g. amino acid or AA) into the peptide chain that grows from the N- to the C-terminus and associated functional group modifications (Sieber and Marahiel 2005). Modules are composed of domains that catalyze the single reaction steps like activation, covalent binding, optional modification of the building blocks, and condensation with the amino acyl or peptidyl group on the neighbouring module (Mootz et al. 2002). At least three domains are necessary for the non-ribosomal production of peptides (see Fig. 1): An adenylation (A) domain, a peptidyl carrier protein (PCP) also denoted as thiolation (T) domain, and a condensation (C) domain (Grünewald and Marahiel 2006). These three domains are called 'core' domains.

The first (*N*-terminal) module ('start' module) of an NRPS often lacks the C domain, and the last (*C*-terminal) module ('termination' module) usually comprises a thioesterase (TE) domain (Sieber and Marahiel 2005). The TE domain is usually responsible for the release of linear (transfer to a water molecule), cyclic or

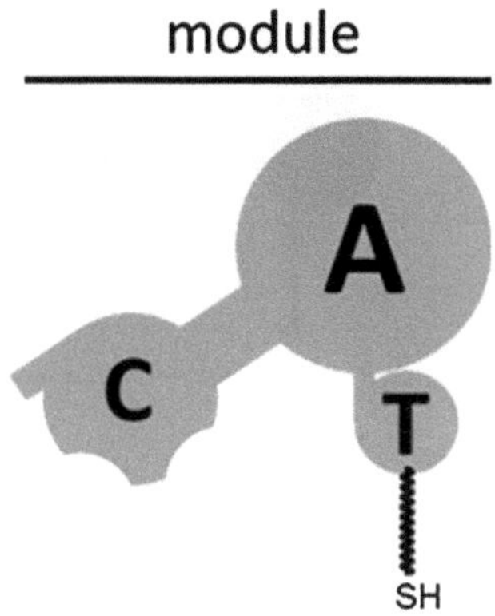

Fig. 1 A canonical NRPS module showing the core domains. Condensation (*C*), adenylation (*A*) and thiolation (*T*) domains are shown, the latter in its *holo*-form with the phosphopantetheinyl arm (*wavy line*) attached to a conserved serine residue

branched cyclic peptides (amide or ester linkage). In addition to these 'standard' domains (*C*, *A*, *T*, TE), a cyclization (Cy) domain instead of a *C* domain and a terminal condensation (C_{term}) domain in place of a TE domain may also be present. Further, modification domains like an epimerization (*E*) domain, N-methylation (MT) domain or oxidation (Ox) domain can also be part of such modules.

The reactions catalyzed by standard NRPS domains are now well understood (Marahiel 2016; Sieber and Marahiel 2005) and their essential enzymatic activities are illustrated in Fig. 2 (Mootz et al. 2002). These activities reside in the adenylation (*A*), thiolation (*T*) and condensation (*C*) domains. First, AAs are activated through the activity of an A domain (Fig. 2a). The energy derived from ATP hydrolysis is used to form an aminoacyl-adenylate intermediate. Then the AA is loaded on the thiol of the pantetheine cofactor of the T domain (Fig. 2b) (Gulick 2009). Consecutively activated AAs on T domains are then joined by a C domain that catalyzes peptide bond formation and transfers the upstream AA or peptide to the downstream substrate (Fig. 2c).

In most cases the number of NRPS modules corresponds directly to the number of AA residues incorporated into the associated peptide and the arrangement of the modules directly follows the peptides' primary sequence because peptide synthesis proceeds collinear in an *N*- to *C*- terminal direction (Mootz et al. 2002). Such biosynthetic templates are called 'linear' or 'type A' NRPS (Fig. 3). Examples for linear NRPSs are the tyrocidine (Mootz and Marahiel 1997), surfactin (Cosmina et al. 1993) and GameXPeptide (Nollmann et al. 2015a) assembling NRPS.

In contrast to type A NRPSs, there are also 'iterative' or 'type B' and 'nonlinear' or 'type C' NRPSs (Mootz et al. 2002) (Fig. 5). Iterative NRPSs (e.g. enniatin, enterobactin) use their modules or domains more than once in the assembly of a single peptide. This strategy is employed to build up peptide chains that consist of shorter sequences that are repeated. Nonlinear NRPSs, such as vibriobactin, yersiniabactin and fungisporin, deviate in their domain organization from the standard $(C\text{-}A\text{-}T)_n$ module architecture (Mootz et al. 2002). They are characterized by at least one unusual arrangement of the core domains *C*, *A* and *T*, A deviation

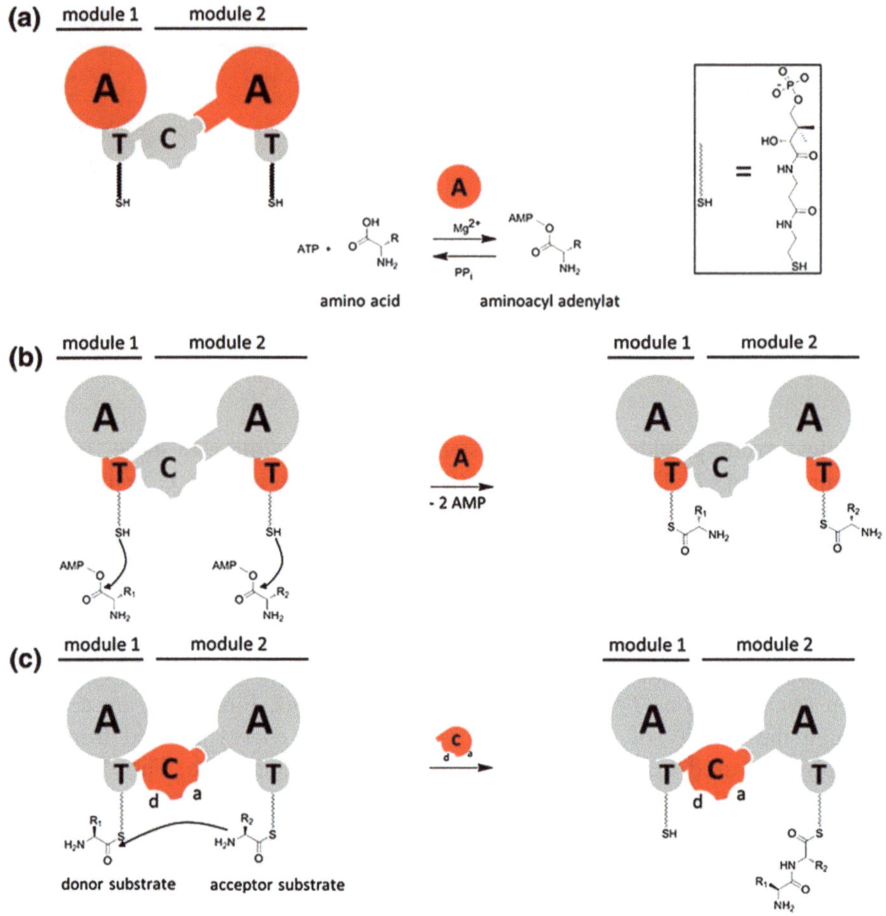

Fig. 2 Essential enzymatic activities in non-ribosomal peptide synthases or NRPS. Active domains are shown in red. **a** Substrate recognition and activation by the A domain. **b** Covalent attachment of the activated aminoacyl adenylate onto the T domain bound 4'Ppan cofactor. **c** Peptide elongation. The *C* domain catalyzes an attack of the nucleophilic amine of the acceptor substrate onto the electrophilic thioester of the donor substrate. *C* domain's acceptor site is indicated by '*a*', the donor site by '*d*'

from the module arrangement $(C\text{-}A\text{-}T)_n$ of linear NRPSs often goes along with nonlinear peptide products that result from unusual internal cyclizations (e.g. bleomycin) or branch point syntheses (e.g. mycobactin). Consequently, the peptide sequence does not represent the linear module and domain arrangement of the biosynthetic template itself.

Two further cases must be considered regarding the role of inter-domain recognition in biosynthetic templates: (I) when domains are next to each other on the same polypeptide chain (domains acting in *cis*) or (II) when they belong to two distinct

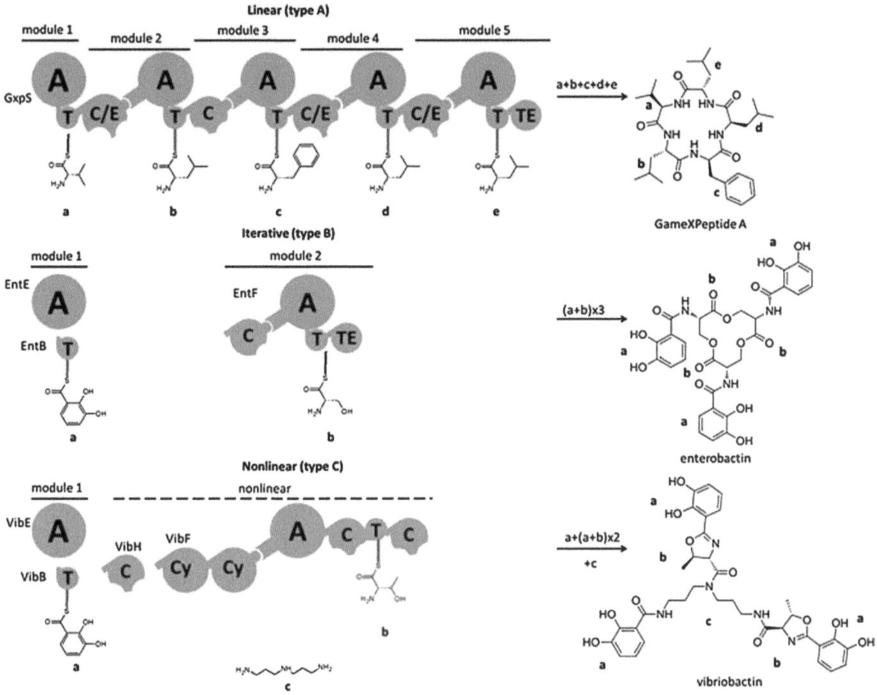

Fig. 3 Biosynthetic strategies of non-ribosomal peptide synthases or NRPS. Small letters (*a*–*e*) indicate the amino acids or building blocks and their connectivity

proteins (domains acting in *trans*), as NRPSs can be composed of several polypeptide chains (Lautru and Challis 2004). In the latter context, correct protein–protein recognition between NRPSs acting *in trans* is crucial to prevent unspecific interactions with non-partner enzymes and to guarantee an effective biosynthesis of the desired product, e.g. the tyrocidine biosynthetic systems consists of three distinct NRPSs (TycA, TycB and TycC) (Hahn and Stachelhaus 2004, 2006). The intermolecular communication within multienzyme NRPS complexes relies on the coordinated interplay of donor and acceptor communication-mediating (COM) domains (Chiocchini et al. 2006). In various studies, it has been shown that matching pairs of donor and acceptor domains promote the correct positioning of enzymes within multi-enzyme complexes and the selective translocation of intermediates between adjacent synthetases (Hahn and Stachelhaus 2004, 2006). COM domains are composed of 15–30 AA located at the *C*- and *N*-terminal end of the corresponding polypetide chains and possess α-helical structures that provide the interfaces for the selective differentiation between partner and non-partner NRPSs. In the specific case of multifunctional but single-protein NRPSs, when domains on the same polypeptide chain are interconnected through short stretches of amino acid residues, domain communication is acting in *cis* via defined catalytically inactive linker regions

($\sim$15 AA) (Lautru and Challis 2004). Studies on many systems have therefore provided compelling evidence that linkers are more than 'simple' covalent connectors (Gokhale and Khosla 2000).

3 Overview on Biosynthesis Gene Clusters and Natural Products in Different *Photorhabdus* Strains

A detailed analysis of the genome sequence of seven *Photorhabdus* strains (*P. luminescens* TTO1, *P. luminescens* PB45.5, *P. asymbiotica* ATCC 43949, *P. asymbiotica* subsp. australis PB68.1, *P. temperata* subsp. *thracensis* DSM 15199, *P. temperata* subsp. *temperata* M1021 and *P. temperata* subsp. *khani* NC19) revealed the presence of up to 22 different BGCs in each of these strains (Tobias et al. 2016b). The main BGCs were NRPS but PKS-NRPS hybrids and other BGC classes were also detected. Interestingly, only a few BGC were found common to all strains: the BGCs for isopropylstilbene (IPS)/Dialkylresorcinol (DAR) and rhabduscin biosynthesis. Rhabduscin (Fig. 4) is an isonitrile NP, a type that is usually rare in nature, and acts as a major player in overcoming the insect immune system via inhibition of the insect enzyme phenoloxidase (Crawford et al. 2012). Phenoloxidase is important in the insect's blood clotting reaction which involves melanisation and is discussed elsewhere in this volume in the chapter on insect immunity by Eleftherianos et al. So far it is one of the few NP classes found in both *Photorhabdus* and *Xenorhabdus* bacteria, as well as in all of their genomes analysed to date (our unpublished data), indicating its important biological function.

DAR with IPS that can be regarded as a special DAR derivative (Fig. 4) and has been shown to be important for development of the nematode symbione (Joyce et al. 2008). Additionally, IPS shows potent antibiotic activity against other bacteria is cytotoxic against eukaryotic cell lines probably due to its inhibition of the soluble epoxide hydrolase (sEH) (Buscató et al. 2013) and DARs have been shown to be signalling compounds involved in quorum sensing in the insect and human pathogen *P. asymbiotica* (Brameyer et al. 2015) (see also the chapter by Ralf Heermann in this volume). Biochemically, DARs are derived by a consecutive Claisen condensation and Michael addition of a β-keto- and an α,β-unsaturated-acyl thioester, the latter derived from elongation of phenylalanine derived cinnamoyl-CoA, resulting in the formation of a cyclohexanedione (CHD) compound that can be oxidized leading to the formation of DARs (Fig. 5) (Fuchs et al. 2013). When α,β-unsaturated-acyl thioesters from the fatty acid biosynthesis or degradation are used, additional DARs are formed as detected in P. *asymbiotica* PB68.1. IPS can be further oxidized to give epoxystilbene showing antibiotic and cytotoxic activity and was found in *P. luminescens* infected larvae of the Greater waxmoth *Galleria mellonella* (Hu et al. 2006). However, it is also produced when *P. luminescens* grows in standard Luria Broth (LB) medium and several NPs derived from epoxide opening and additional oxidations have been found (Kontnik et al. 2010). DARs in general seem to be a widespread class of NPs since the corresponding BGCs are found in several different

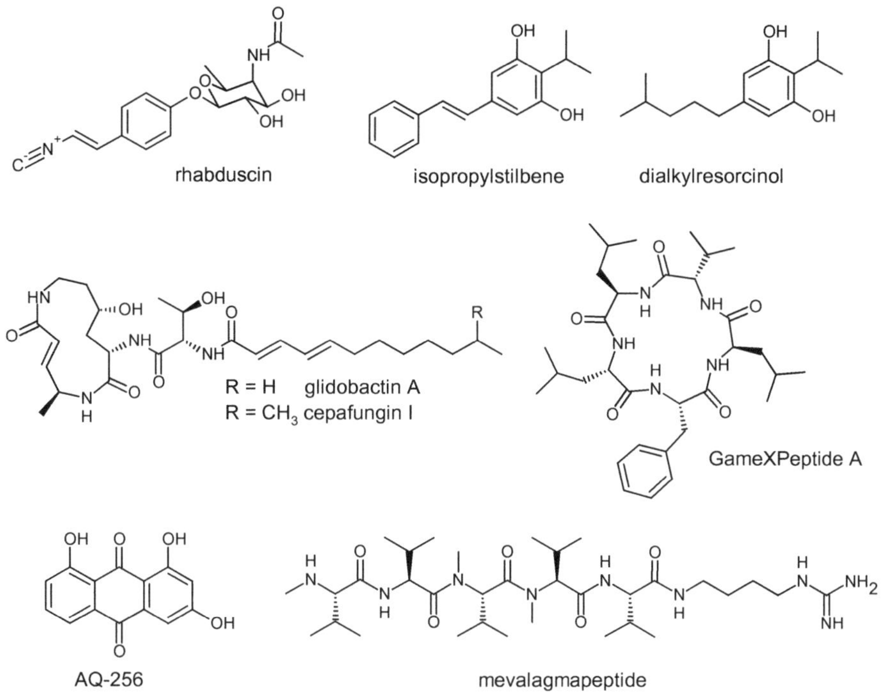

Fig. 4 Conserved natural products identified from *Photorhabdus* bacteria

bacteria, several of which are pathogenic to man (Brameyer et al. 2015). Thus, it might be possible that these bacteria also use these NPs as signalling compounds, which would make them a widespread and medically highly relevant NP class that warrants further investigation (Schöner et al. 2015).

Glidobactins (Fig. 4) have been identified from *P. asymbiotica* infected crickets (Theodore et al. 2012), from *P. luminescens* grown in low sodium chloride (Stein et al. 2012) and from heterologous expression of the encoding BGC in *E. coli* (Fu et al. 2012b). The true natural product and major derivative might be cepafungin I, showing an iso-branched acyl moiety. Glidobactin and cepafungin derivatives are potent proteasome inhibitors that could be involved in protecting the insect cadaver against fungi or other invading saprophytes. It's BGC has been found in all *P. luminescens* and *P. asymbiotica* strains examined to date but not in strain DSM 15199 from the P. *temperata* strains, suggesting that its production may be limited to these groups.

The cyclic pentapeptide GameXPeptides (Bode et al. 2012) have been identified in strain TTO1 and its encoding BGC is found in all but one *P. temperata* strain examined to date (Nollmann et al. 2015a). In *P. luminescens* strain TTO1 infected insects derivatives with p-aminophenylalanine (PAPA) or p-methylaminophenylalanine (PMAPA) have been found instead of the usual phenylalanine. Since the titer of these

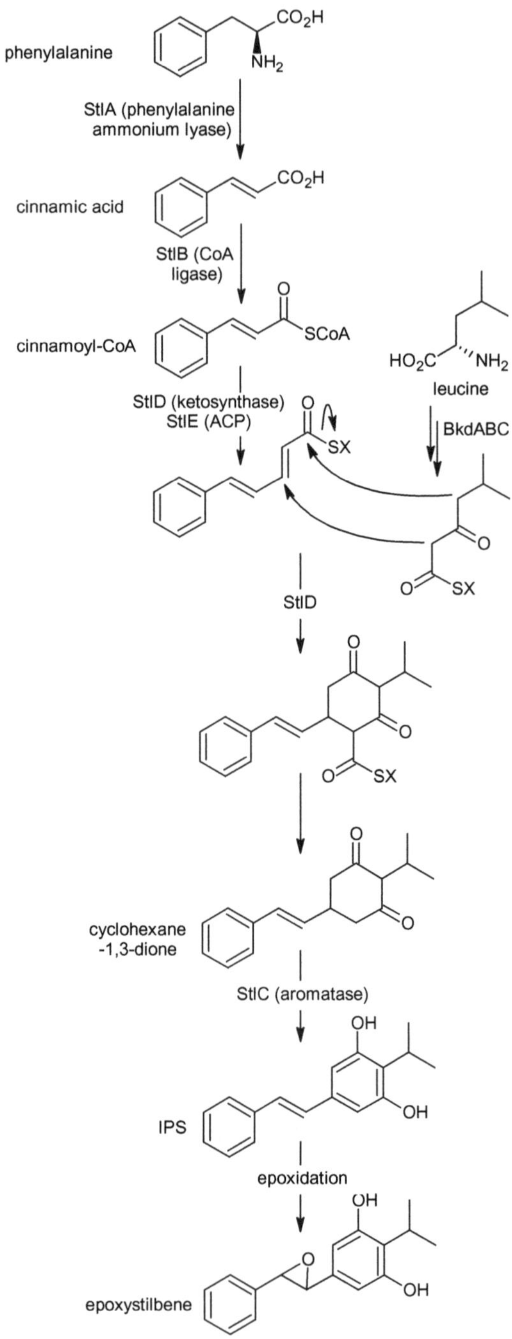

Fig. 5 Biosynthesis of isopropylstilbene (IPS)

derivatives was much higher, they might actually represent the true natural products produced in a normal infection (Nollmann et al. 2015a). Unfortunately, no biological function has been found for these compounds, despite the fact that they are also widespread in *Xenorhabdus* strains.

Anthraquinones (AQ) (Li et al. 1995) are responsible for the orange-red pigmentation of *P. luminescens* and *P. temperata* and its BGC has been identified in these strains but not in *P. asmybiotica* which often lacks this orange pigmentation. They are derived from a type II PKS system (Brachmann et al. 2007) that is widespread in Gram-positive *Streptomyces* bacteria but has so far only been found twice in Gram-negative bacteria. The functions of the AQs are unknown but it has been speculated that they act as ant or bird deterrent factors since similar functions have been observed for AQs in other organisms. However, this has not yet been proven experimentally for AQs from *Photorhabdus*.

The rhabdopeptide encoding BGC that has been shown in *P. luminescens* to be responsible for the production of mevalagmapeptides (Bode et al. 2012, 2015a) is present in five of the seven analysed fully sequenced *Photorhabdus* genomes (see above) and only missing in *P. asymbiotica* ATCC 43949 and *P. temperata* NC19. Rhabdopeptides are very common in *Xenorhabdus* and *Photorhabdus* strains and some derivatives show both cytotoxic and protozoal activity and thus they might also be involved in protection of the insect cadaver from grazers and saprophytes in the soil. There are a few other BGCs for which no NP has been identified yet, including; two NRPSs, a monomodular type I PKS and additional bacteriocins that are widespread but not common to all strains (Tobias et al. 2016b). Besides these conserved clusters found in several of the analysed strains, additional BGCs are found in only one or two strains.

Kolossin (Fig. 6) is a pentadecapeptide produced by TTO1 (Bode et al. 2015b) that is derived from one of the largest bacterial NRPS with a molecular mass of 1.8 MDa. The Kol NRPS has 46 distinct domains and the respective gene is 49 kbp in length thus alone being 1 % of the bacterial genome. Kolossin could only be obtained from activation of the promoter as discussed below and no function could be assigned yet for this rather simple D/L-peptide made by such a gigantic enzymatic machinery.

Photopyrones (PPY) were found in TTO1 and DSM 15199 (Brachmann et al. 2013) and together with the DARs they are the second example of bacterial quorum sensing molecules different to standard acylhomoserine lactones (AHL) (see also the chapter by Ralf Heermann in this volume). PPYs are made from the same β-keto-thioester also required in DAR and IPS biosynthesis (Fig. 5) but here a homodimeric ketosynthase fuses this thioester to an acyl-thioester resulting in α-pyrone formation (Kresovic et al. 2015). Crucial for the activation of the acylthioester is a glutamine residue in the ketosynthase that enters the active site of the second monomer. Although other α-pyrones are known in NP like corallopyronin or myxopyronine the ketosynthases involved in their biosynthesis do not require the glutamine activation (Erol et al. 2010; Sucipto et al. 2013).

The ink blue coloured pigment indigoidine (5,5′-diamino-4,4′-dihydroxy-3,3′-diazadiphenoquinone-(2,2′)) is derived from a monomodular NRPS found in half of

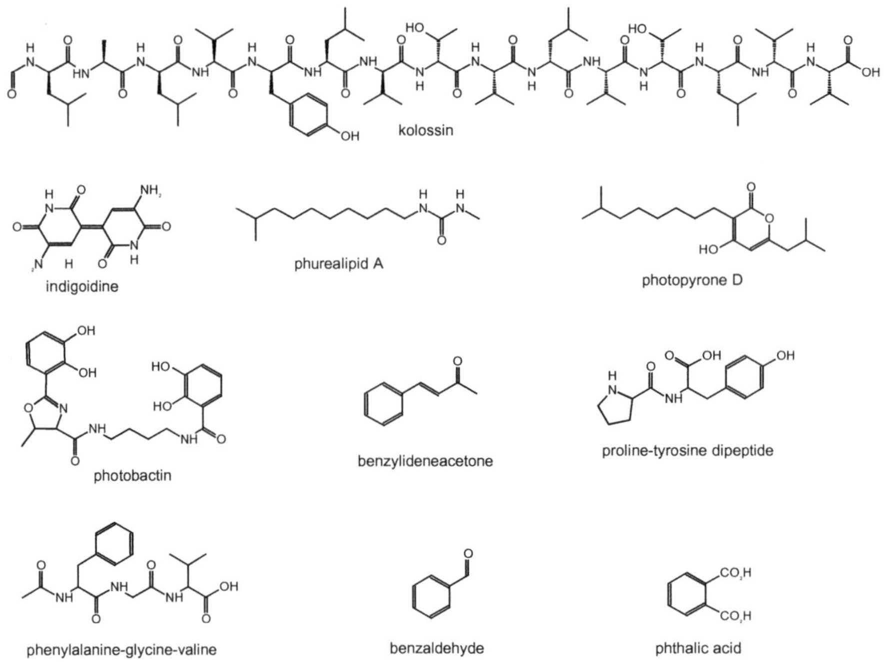

Fig. 6 The structures of a range of natural products from *Photorhabdus* strains that appear to be strain or group specific in their distribution

the analysed strains. The compound itself was identified in *P. luminescens* strain TTO1 using a promoter exchange approach (see below) but no indigoidine has been detected under any growth condition tested (Brachmann et al. 2012). However, from the biophysical properties of indigoidine one might assume a protective function against reactive oxygen species (ROS) or UV light.

Phurealipids have been found in several *P. luminescens* strains, rarely in *P. temperata* and *Xenorhabdus* but are absent from *P. asymbiotica* (Nollmann et al. 2015b). They have been identified as inhibitors of the juvenile hormone epoxide hydrolase that is required for normal development (moulting) and immunity in insects. Phurealipids are derived from the fatty acid metabolism where they are reduced as aldehydes that are transaminated, carbamoylated and methylated. A carbamoyltransferase and a methyltransferase have been identified in the *P. luminscens* strain TTO1 genome but a candidate for the aminotransferase is still missing.

Ciche and co-workers isolated a catecholate siderophore, named photobactin, from *P. luminescens* that is also derived from a NRPS pathway. Due to its role as a siderophore, sequestering and transferring Fe^{3+} into the bacterial cells, photobactin was required for growth under ion-limited conditions (Ciche et al. 2003). However, photobactin is not needed for *P. luminescens* to support the growth and the reproduction of its nematode host. However, purified photobactin was shown to have antibiotic activity, suggesting it may play a role in inhibiting competing bacteria in the insect cadaver. Siderophore encoding BGCs are found in all sequenced *Photorhabdus* strains most of them containing NRPS independent biosynthesis pathways.

In 2004, Ji et al. identified benzylideneacetone (Fig. 6) from *X. nematophila* being active against some Gram-negative bacteria (Ji et al. 2004). From its structure it might be derived from a cinnamic acid extended by one malonyl-CoA followed by decarboxylation. However, this might suggest the presence of a phenylalanine ammonium lyase and CoA-ligase in *X. nematophila* that has not been identified in the available genome sequences so far (our unpublished data). In addition to benzylideneacetone, a linear proline-tyrosine dipeptide and an acetylated phenylalanine-glycine-valine tripeptide were both isolated and they were also shown to be phospholipase A2 inhibitors. These three compounds could also be isolated from *P. temperata* culture broth. Additionally, *X. nematophila* can produce four additional phospholipase A2 inhibitors (indole, oxidole, *cyclo*-proline-tyrosine dipeptide and p-hydroxyphenyl propionic acid) (Seo et al. 2012). Phospholipase A2 is crucial for the insect's immune response and thus needs to be overcome in order to colonize the insect.

Two very small compounds, benzaldehyde (Ullah et al. 2015) and phthalic acid (Ullah et al. 2014), were isolated from *P. temperata*. Benzaldehyde possessed antioxidant, insecticidal and antimicrobial activities. Moreover, phthalic acid has the capacity to inhibit phenoloxidase with a consequence to suppress the insect's immune defence.

Derzelle and co-workers have identified the putative gene cluster of a carbapenem-like antibiotic in *P. luminescens* (Derzelle et al. 2002) and two types of carbapenem BGCs have been identified in four of the seven genomes annotated to date. However, their exact structure is not known. Carbapenem antibiotics are members of the β-lactam family of antibiotics, which are now the most important class of antibiotics for clinical use and their biosynthesis and regulation has been extensively studied in *Serratia* (Coulthurst et al. 2005).

4 Tools for the Identification of Natural Products

In order to identify NPs from *Photorhabdus* or any other organism one can cultivate the strain of interest using several different growth conditions summarized in the OSMAC (One Strain Many Compounds) approach (Bode et al. 2002) that often results in the production of different NP classes at different conditions varying media composition, aeration, pH, salt concentration and even co-cultivation with other organisms to mimic the natural condition at which the respective NP is produced. One can also try a range of molecular techniques to enhance the production of the NP in order to enhance its detection and purification. Each of these will now be discussed below.

4.1 Promoter Exchange

One can also use the BGC information obtained from whole genome sequencing for a 'promoter exchange' approach. Since the natural promoters might have unknown modes of regulation via specific transcription factors or other unknown mechanisms, they can be replaced by strong constitutive or inducible promoters (Biggins et al. 2014, 2011; Bode et al. 2015a). The latter approach is especially beneficial since it allows to compare the wild-type strain with the induced (usually overproduction of the NP derived from BGC overexpression) and the non-induced promoter exchange mutant (usually a non-producer due to the missing expression) (Fig. 7). This simple approach has also been used widely by others but was especially useful in *Photorhabdus* and *Xenorhabdus* with the arabinose inducible

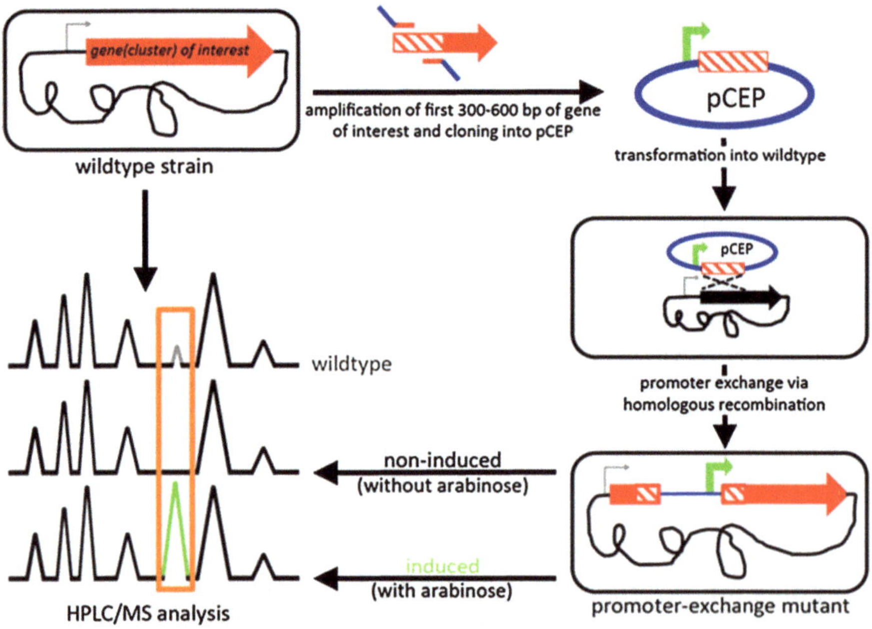

Fig. 7 Schematic representation of the promoter exchange approach used in *Photorhabdus* and *Xenorhabdus* (Bode et al. 2015a). The start (300–600 bp) of the gene of interest is amplified by PCR and cloned into the cluster expression plasmid pCEP. Following transformation into *Photorhabdus*, the genomic insertion of the non-replicating plasmid results in a promoter exchange mutant in which the expression of the full-length gene is not driven by the natural promoter (grey arrow) but by the introduced inducible promoter (green arrow; here P_{BAD}). The resulting promoter exchange mutant can easily be selected based on the pCEP-encoded resistance gene. The new promoter is tightly controlled and shows no activity without the inducer arabinose. Therefore, the non-induced strain behaves like a knockout mutant (no production of the compound of interest) whereas overexpression of the desired gene is achieved with arabinose resulting in an overproducing mutant (in *green*, relative to the wild-type strain). From the comparison of these three strains the NP can easily correlated to the BGC in the genome

P_{BAD}/AraC system resulting in the production of xenoamicins, GameXPeptides, mevalagmapeptides, the blue pigment indigoidine, the cytotoxic yellow pigment xenorhabdin and kolossin (Bode et al. 2015a, b). The latter is derived from one of the largest bacterial NRPS, named *Kol*, with a molecular weight of 1.8 MDa and 49 catalytically active domains covering almost 1 % of the *Photorhabdus* chromosome. *Kol* is only expressed in a promoter exchange mutant and no natural condition has been found so far for its production (Bode et al. 2015b). The power of the method is also evident from the fact that for xenoamicin productin in *Xenorhabdus doucetiae* the production titer is well beyond the solubility of these depsipeptides and white crystals can therefore be observed in the expression culture. For GameXPeptides the 10-fold higher production titer compared to the wild-type level was accompanied by the occurrence of the linear versions of the usually cyclic peptides (Bode et al. 2015a).

4.2 Manipulation of Regulatory Proteins

Another approach is the manipulation of regulatory proteins involved in NP production. This approach was especially useful for fungi that often encode a specific transcriptional regulator as part of their NP BGC (Scharf and Brakhage 2013). In *Photorhabdus* it was shown that HexA (see chapter by David Clarke also in this volume) and the two-component system BarA-UvrY (see chapter by David Clarke in this volume) are involved in NP regulation. Previous research has concentrated on the UV active NP isopropylstilbene and anthraquinones that are both influenced by these regulators. A detailed analysis of a *hexA* deletion and overexpression mutant using the promoter exchange approach described above revealed that in a *hexA* deletion GameXPeptides and mevalagmapeptides are overproduced while some phurealipids are strongly decreased (our unpublished data). Similar analyses with the global regulators LeuO and Lrp showed that they also affect NP production. LeuO is a LysR type transcription factor (Hernandez-Lucas and Calva 2012) involved in virulence of *Vibrio cholerae* and Lrp (Brinkman et al. 2003) has been described as a global regulator affecting mutualism with the nematode and pathogenicity against the insect in *Xenorhabdus nematophila* (Cowles et al. 2007). We were able to show that Lrp is required for the production of rhabdopeptides in *X. nematophila* (our unpublished data) that show insecticidal activity. In *P. luminescens* TT01 *lrp* deletion increases mevalagmapeptide production but decreases phurealipid production while overexpression restored wild-type production levels but led to increased desmethylphurealipid levels. Phurealipid biosynthesis is not encoded in an operon and currently only two steps in their biosynthesis encoded in different genomic loci are known. However, from these results it is quite obvious that both genes are differentially regulated leading to the observed production phenotype using *hexA*, *lrp* or *leuO* mutants.

The most dramatic effect of all regulatory mutants analysed in our group was that of a *hfq* deletion mutant in *P. luminescens*. Hfq is a RNA chaperone mediating interaction between small regulatory RNA and mRNA that is involved in the regulation of virulence in several different bacteria (De Lay et al. 2013; Vogel and Luisi 2011). In *P. luminescens* a Δ*hfq* strain did not produce any known natural product (Fig. 8). In fact the strain looked more like an *E. coli* strain with essentially no NP produced whereas genetic complementation with a plasmid encoded *hfq* restored NP production although not to the wild-type level for all NP classes

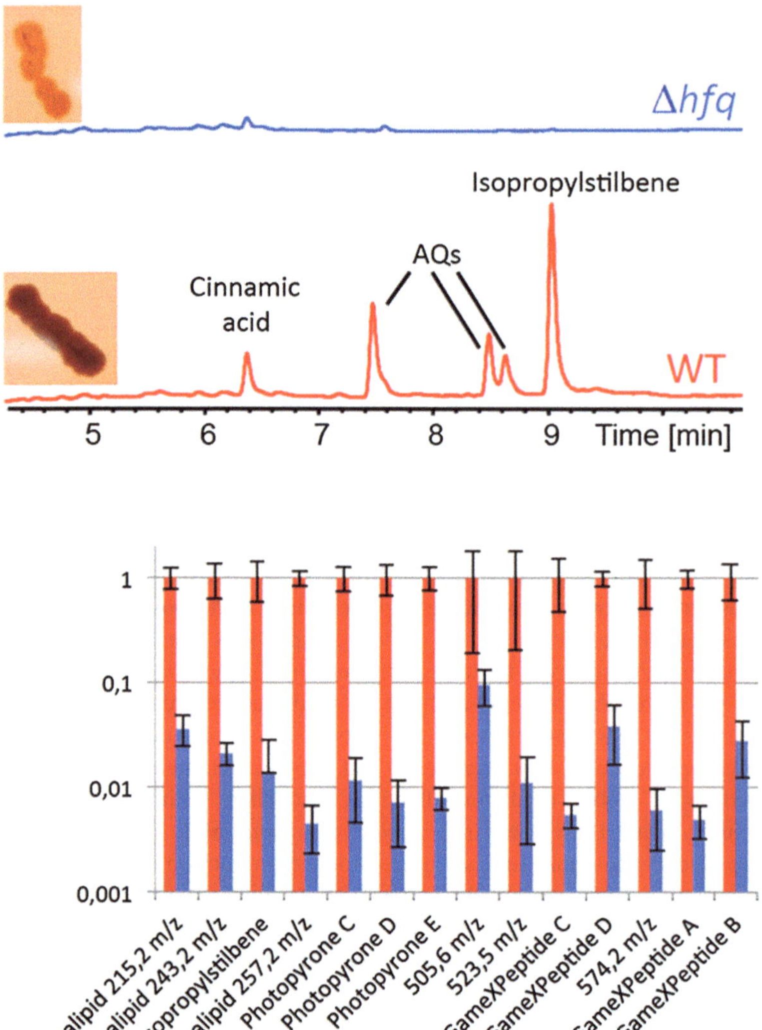

Fig. 8 Comparison between *P. luminescens* wild-type and a Δ*hfq* strain. HPLC/UV chromatograms and colony colour are shown at the top, HPLC/MS quantification of known (with names) and unknown NP (with mass in formation in m/z) at the bottom

(Tobias et al. 2016a). Interestingly, while no decrease in insect pathogenicity was observed for a Δ*hfq* strain, no nematode recovery at all was observed when *Heterorhabditis bacteriophora* was developed on the Δ*hfq* strain. This might indicate no crucial role of NP in insect pathogenesis but an even more important function in the symbiosis with the nematode as suggested previously for iso-propylstilbene (Joyce et al. 2008). Here the next step clearly is the identification of the underlying regulatory principles and especially the small RNA(s) that cause this severe phenotype.

4.3 Heterologous Expression

In general heterologous production of BGCs from *Photorhabdus* and *Xenorhabdus* works well in *E. coli* and direct cloning and yeast based cloning have both applied successfully in the past (Crawford et al. 2012; Fu et al. 2012a; Schimming et al. 2014). However, the promoter exchange approach as well as the manipulation of regulatory proteins might be superior to the heterologous expression of the BGC of interest since all precursors needed for the production of the 'real' NP are present in the original producer while the heterologous host might not have these precursors. Even simple building blocks like amino or fatty acids can be different and for *P. luminescens* it has been shown that heterologous expression of the GameXPeptide producing NRPS GxpS in *E. coli* led to the production of the usual GameXPeptides A-D while overexpression of *gxpS* results in derivatives carrying p-amino- or p-aminomethyl phenylalanine (Nollmann et al. 2015a). These derivatives are the major derivatives produced in insects since this activates a BGC for the production of these unusual phenylalanine derivatives. Similarly, heterologous expression of the glidobactin producing BGC resulted in the production of glidobactin A in *E. coli* (Dudnik et al. 2013; Fu et al. 2012a) while the major derivative in *P. luminescens* is in fact cepafungin (Stein et al. 2012; Theodore et al. 2012) having an iso-branched fatty acid. However, *E. coli* is not able to produce such fatty acids.

4.4 Chemical Tools

Beyond these molecular methods, one can also use further analytical chemistry to identify and characterize NP. Here the major breakthrough was the development of mass spectrometry to allow the very sensitive detection of NPs and even the visualization of NP production in symbiotic systems or other interactions. Mass spectrometry also allows the structural elucidation of NPs without their actual isolation and in combination with labelling experiments even the configuration of a peptide NP can be assigned as demonstrated for the GameXPeptides and kolossin from *Photorhabdus* (Bode et al. 2012, 2015b).

Isotopic labelling can be regarded as a special form of precursor-directed biosynthesis that allows the identification of NP building blocks and even the generation of NP derivatives with modified biological activity. Even more useful would be the potential incorporation of a chemical label that not only allows for the rapid localization of any given NP but can be also used for NP enrichment. One such label is the azide and/or alkyne group as these groups are small and are often well tolerated by the biosynthesis machinery of the cell. They can also react efficiently and specifically without complicated workup in a Huisgen 1,3-dipolar cycloaddition (Kolb et al. 2001). The proven applications of this prototype of a so-called 'click reaction' range from inorganic and organic synthesis to in vitro labelling. The true milestone in development and optimization of this technique was achieved by Bertozzi and co-workers via the rediscovery of ring-strained cyclooctynes as reaction partners for azides without the need of any possibly cytotoxic copper catalysts, a process coined the strain-promoted azide-alkyne cycloaddition (Fig. 9) (Jewett and Bertozzi 2010).

This discovery opened up the now very popular field of 'bioorthogonal' chemistry, where two reaction partners (e.g. an alkyne and a cyclooctyne) can react even within a live cell without substantially altering or harming it. This is possible because in most organisms there is no natural reaction partner for either a cyclooctyne or an azide is present. With the azido group also being a rather small functional group, roughly the size of an ethyl group, not featuring a noteworthy polarity or reactivity, it is the perfect label for in vivo investigations.

Since fatty acids with ω-alkyne or –azide groups have been used previously for the detection and enrichment of lipoproteins (Hang and Linder 2011; Hang et al. 2007), we tested the incorporation of ω-azide fatty acids (AFA) in NP produced by *Photorhabdus* and *Xenorhabdus* and indeed we could detect such derivatives after reaction with an cleavable azide reactive resin (CARR) and the subsequent enrichment of triazole (Fig. 10) (Pérez et al. 2016). The advantage of this method is the gain in MS sensitivity through the very efficient ionization of the triazole product formed as well as the characteristic fragmentation pattern of the clicked derivative that enabled the reliable identification of the clicked products and thus the original azide containing NP.

In *Photorhabdus* AFA-based CARR led to the identification of phurealipids, whereas in *X. doucetiae* phenylethylamides were detected. Applying

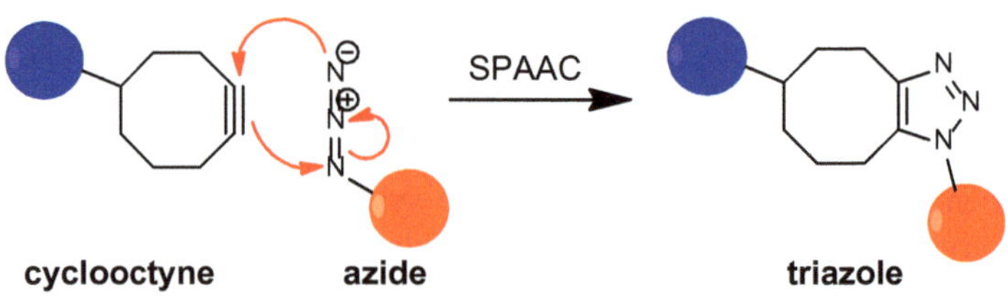

Fig. 9 General mechanism of the strain-promoted alkyne-azide cycloaddition (SPAAC) with octynes leading to a very stable triazole derivative

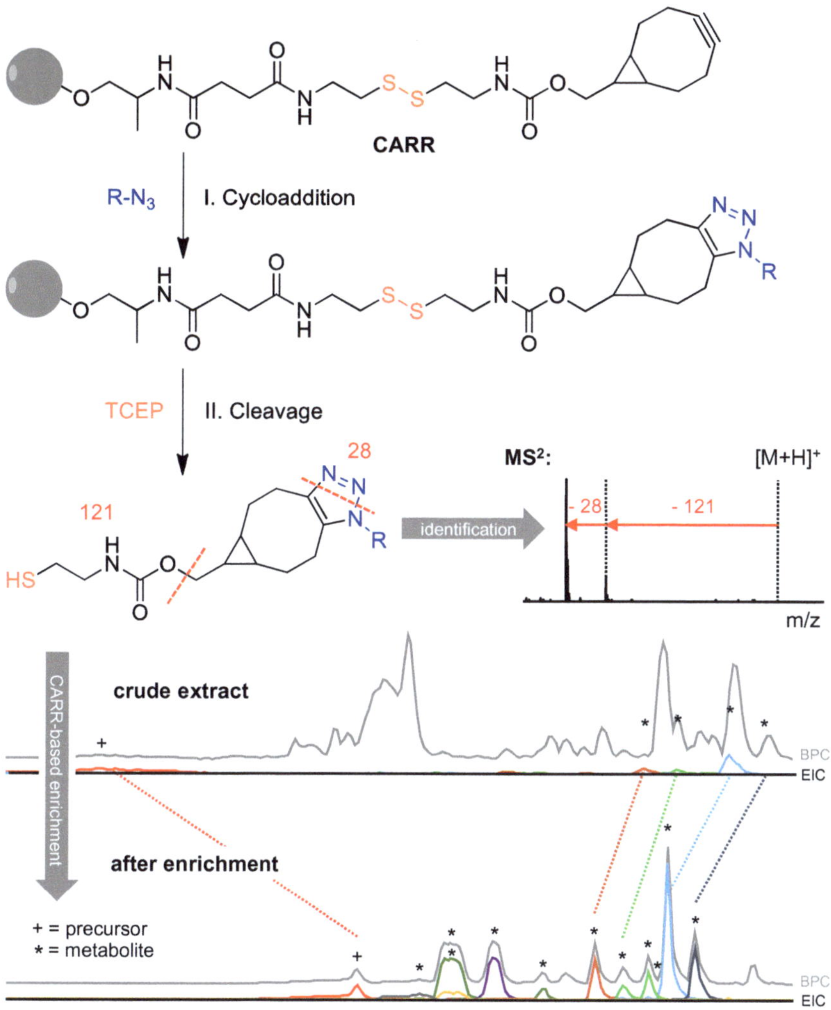

Fig. 10 Overview on azide enrichment using cleavable azide reactive resin (CARR) showing the typical fragmentation of triazole products formed as well as a typical example from a bacterial extract before and after CARR enrichment indicating the purification of only CARR-reacted azides (see BPC and EIC for comparison). BPC (base peak chromatogram = all ions are shown), EIC (extracted ion chromatogram = only ion traces for specific compounds are shown), TCEP (tris (2-carboxyethyl)phosphine))

p-azidophenylalanine based CARR in *P. luminescens* GameXPeptide A, cinnamic acid and cinnamoyl-phenylalanine (all in azidated form) have been identified (Pérez et al. 2016). The latter represents a new class of NP from *P. luminescens* of which the non-azidated form has not been identified yet. Interestingly, similar compounds were also identified in *X. szentirmaii* during p-azidophenylalanine-based CARR (Fig. 11).

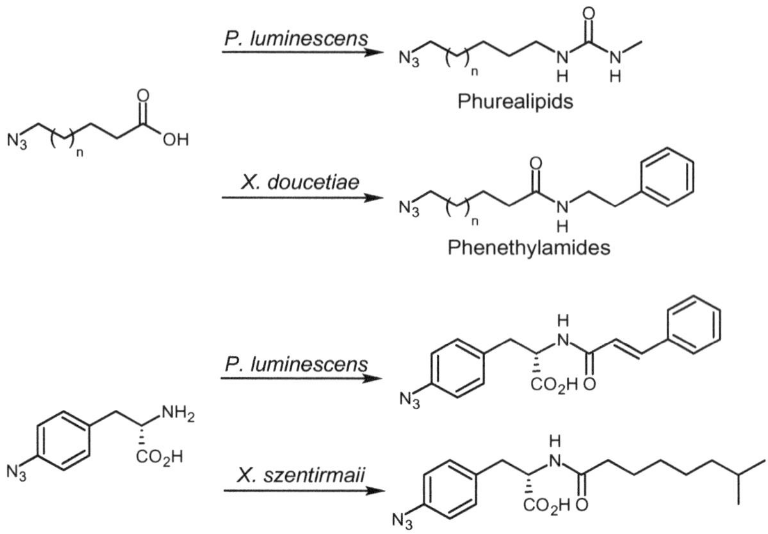

Fig. 11 Identified NP using AFA (top) and p-azidophenylalanine (bottom) based CARR

The fact that also azidated peptides like GameXPeptide or szentiamide from *X.* szentirmaii or xenortide from *X. nematophila* have been detected resulting from the substitution of phenylalanine by p-azidophenylalanine indicates the flexibility of the responsible NRPS to incorporate even these non-natural amino acids. Such peptides might be of great interest for the drug discovery process since they can be further modified chemically after their isolation using the same click chemistry used in the CARR enrichment.

5 Conclusions

During the last 10 years NP research within entomopathogenic bacteria has made huge advances due to the diligent work of several laboratories. However, there is still a lot of information missing that requires further research. One goal for the future must be to develop more tools for the rapid target identification of the various NPs produced. We need these tools to answer the wide range outstanding biological questions that remain. For example, to what enzyme or receptor do these NPs bind and what is the resulting phenotype in the insect, the nematode, the original producer or food competitors? To achieve this goal classical mode of action studies must be performed (Schmitt et al. 2015), however, these are often long and technically challenging studies. Reactive NP derivatives can be synthesized that can covalently bind to the NP target and in case a handle like an azide is also present in these NP derivatives one can use this to isolate the NP-target complex as previously shown.

Additionally, if one assumes that all *Photorhabdus* strains are adapted to the same ecological niche being mutualistic to nematodes and pathogenic towards insects. Thus, they might require a certain set of functionally similar chemical tools to fulfil these similar ecological functions. Therefore, one might predict NPs that are functionally similar but can be chemically different and might be derived from different BGCs in strains that do not have a NP with a known functionality. Examples of this strain specificity of NP production might be the *P. temperata* strains lacking glidobactins. In which case, here it would be interesting to look for other proteasome inhibitors that might have chemically new scaffolds.

Regarding proteasome inhibitors, and other toxic NP compounds, it is still a mystery why the nematode host can survive in the dead insect containing such large amounts of these toxic compounds. Once the targets of these NP in the insect are known one must investigate whether they are different in the nematode host or how it can protect itself from their activity. A resistance against toxic NPs from the bacterial symbiont can also be a protection mechanism against other nematodes and might ensure the specificity of the symbiosis by killing the wrong nematode.

Finally, with the already developed molecular tools for manipulation of the bacteria and the first available genome sequences for the entomopathogenic nematodes (Bai et al. 2013; Dillman et al. 2015) that allow also transcriptomics and proteomics one can also start to address the NP function from the hosts side by manipulation of NP regulated signalling pathways or other NP targets. Moreover, similar work can be done in insects where all tools are already available and just need to be used to address the functions of NPs from entomopathogenic bacteria that again can be analysed in depth using 'omics' technologies. It is an exciting time to study these bacteria and once we have understood such a rather simple system of organismic interaction we can also start to look in detail in other and more complex systems involving more interaction partners.

References

Ackermann M (2015) A functional perspective on phenotypic heterogeneity in microorganisms. Nat Publishing Group 13:497–508

Arnison PG, Bibb MJ, Bierbaum G, Bowers AA, Bugni TS, Bulaj G, Camarero JA, Campopiano DJ, Challis GL, Clardy J et al (2012) Ribosomally synthesized and post-translationally modified peptide natural products: overview and recommendations for a universal nomenclature. Nat Prod Rep 30:108–160

Bai X, Adams BJ, Ciche TA, Clifton S, Gaugler R, Kim K-S, Spieth J, Sternberg PW, Wilson RK, Grewal PS (2013) A lover and a fighter: the genome sequence of an entomopathogenic nematode heterorhabditis bacteriophora. PLoS One 8:e69618

Biggins JB, Liu X, Feng Z, Brady SF (2011) Metabolites from the induced expression of cryptic single operons found in the genome of Burkholderia pseudomallei. J Am Chem Soc 133: 1638–1641

Biggins JB, Kang H-S, Ternei MA, DeShazer D, Brady SF (2014) The chemical arsenal of Burkholderia pseudomalleiIs essential for pathogenicity. J Am Chem Soc 136:9484–9490
Bode HB, Müller R (2005) The impact of bacterial genomics on natural product research. Angew Chem Int Ed 44:6828–6846
Bode HB, Bethe B, Höfs R, Zeeck A (2002) Big effects from small changes: possible ways to explore nature's chemical diversity. ChemBioChem 3:619–627
Bode HB, Reimer D, Fuchs SW, Kirchner F, Dauth C, Kegler C, Lorenzen W, Brachmann AO, Grün P (2012) Determination of the absolute configuration of peptide natural products by using stable isotope labeling and mass spectrometry. Chem Eur J 18:2342–2348
Bode E, Brachmann AO, Kegler C, Simsek R, Dauth C, Zhou Q, Kaiser M, Klemmt P, Bode HB (2015a) Simple "on-demand" production of bioactive natural products. ChemBioChem 16:1115–1119
Bode HB, Brachmann AO, Jadhav KB, Seyfarth L, Dauth C, Fuchs SW, Kaiser M, Waterfield NR, Sack H, Heinemann SH et al (2015b) Structure elucidation and activity of kolossin A, the D-/L-Pentadecapeptide product of a giant nonribosomal peptide synthetase. Angew Chem Int Ed Engl 54:10352–10355
Brachmann AO, Joyce SA, Jenke-Kodama H, Schwär G, Clarke DJ, Bode HB (2007) A type II polyketide synthase is responsible for anthraquinone biosynthesis inphotorhabdus luminescens. ChemBioChem 8:1721–1728
Brachmann AO, Kirchner F, Kegler C, Kinski SC, Schmitt I, Bode HB (2012) Triggering the production of the cryptic blue pigment indigoidine from *Photorhabdus luminescens*. J Biotechnol 157:96–99
Brachmann AO, Brameyer S, Kresovic D, Hitkova I, Kopp Y, Manske C, Schubert K, Bode HB, Heermann R (2013) Pyrones as bacterial signaling molecules. Nat Chem Biol 9:573–578
Brameyer S, Kresovic D, Bode HB, Heermann R (2015) Dialkylresorcinols as bacterial signaling molecules. Proc Natl Acad Sci U.S.A. 112:572–577
Brinkman AB, Ettema TJG, De Vos WM, Van Der Oost J (2003) The Lrp family of transcriptional regulators. Mol Microbiol 48:287–294
Buscató EL, Büttner D, Brüggerhoff A, Klingler F-M, Weber J, Scholz B, Živković A, Marschalek R, Stark H, Steinhilber D et al (2013) From a multipotent stilbene to soluble epoxide hydrolase inhibitors with antiproliferative properties. ChemMedChem 8:919–923
Caboche S, Leclere V, Pupin M, Kucherov G, Jacques P (2010) Diversity of monomers in nonribosomal peptides: towards the prediction of origin and biological activity. J Bacteriol 192:5143–5150
Cantley AM, Clardy J (2015) Animals in a bacterial world: opportunities for chemical ecology. Nat Prod Rep 32:888–892
Chiocchini C, Linne U, Stachelhaus T (2006) In vivo biocombinatorial synthesis of lipopeptides by COM domain-mediated reprogramming of the surfactin biosynthetic complex. Chem Biol 13:899–908
Ciche TA, Blackburn M, Carney JR, Ensign JC (2003) Photobactin: a Catechol Siderophore Produced by *Photorhabdus luminescens*, an Entomopathogen Mutually Associated with Heterorhabditis bacteriophora NC1 Nematodes. Appl Environ Microbiol 69:4706–4713
Cosmina P, Rodriguez F, de Ferra F, Grandi G, Perego M, Venema G, van Sinderen D (1993) Sequence and analysis of the genetic locus responsible for surfactin synthesis in Bacillus subtilis. Mol Microbiol 8:821–831
Coulthurst SJ, Barnard AML, Salmond GPC (2005) Regulation and biosynthesis of carbapenem antibiotics in bacteria. Nat Rev Micro 3:295–306
Cowles KN, Cowles CE, Richards GR, Martens EC, Goodrich-Blair H (2007) The global regulator Lrp contributes to mutualism, pathogenesis and phenotypic variation in the bacterium Xenorhabdus nematophila. Cell Microbiol 9:1311–1323
Crawford JM, Portmann C, Zhang X, Roeffaers MBJ, Clardy J (2012) Small molecule perimeter defense in entomopathogenic bacteria. Proc Natl Acad Sci U.S.A. 109:10821–10826
Davies J, Ryan KS (2012) Introducing the parvome: bioactive compounds in the microbial world. ACS Chem Biol 7:252–259

Davies J, Spiegelman GB, Yim G (2006) The world of subinhibitory antibiotic concentrations. Curr Opin Microbiol 9:445–453
De Lay N, Schu DJ, Gottesman S (2013) Bacterial small RNA-based negative regulation: Hfq and its accomplices. J Biol Chem 288:7996–8003
Derzelle S, Duchaud E, Kunst F, Danchin A, Bertin P (2002) Identification, characterization, and regulation of a cluster of genes involved in carbapenem biosynthesis in *Photorhabdus luminescens*. Appl Environ Microbiol 68:3780–3789
Dillman AR, Macchietto M, Porter CF, Rogers A, Williams B, Antoshechkin I, Lee M-M, Goodwin Z, Lu X, Lewis EE et al (2015) Comparative genomics of steinernema reveals deeply conserved gene regulatory networks. Genome Biol 1–21
Dorrestein PC, Mazmanian SK, Knight R (2014) Finding the missing links among metabolites, microbes, and the host. Immunity 40:824–832
Dudnik A, Bigler L, Dudler R (2013) Heterologous expression of a *Photorhabdus luminescens* syrbactin-like gene cluster results in production of the potent proteasome inhibitor glidobactin A. Microbiol Res 168:73–76
Erol Ö, Schäberle TF, Schmitz A, Rachid S, Gurgui C, Omari, EM, Lohr F, Kehraus S, Piel J, Müller R et al (2010) Biosynthesis of the myxobacterial antibiotic corallopyronin A. ChemBioChem 11:1253–1265
Felnagle EA, Jackson EE, Chan YA, Podevels AM, Berti AD, McMahon MD, Thomas MG (2008) Nonribosomal peptide synthetases involved in the production of medically relevant natural products. Mol Pharm 5:191–211
Fu J, Bian X, Hu S, Wang H, Huang F, Seibert PM, Plaza A, Xia L, Müller R, Stewart AF et al (2012a) Full-length RecE enhances linear-linear homologous recombination and facilitates direct cloning for bioprospecting. Nat Biotechnol 30:440–446
Fu J, Bian X, Hu S, Wang H, Huang F, Seibert PM, Plaza A, Xia L, Stewart AF, ller RMU et al (2012b) Full-length RecE enhances linear-linear homologous recombination and facilitates direct cloning for bioprospecting. Nat Biotechnol 30:440–446
Fuchs SW, Bozhüyük KAJ, Kresovic D, Grundmann F, Dill V, Brachmann AO, Waterfield NR, Bode HB (2013) Formation of 1,3-cyclohexanediones and resorcinols catalyzed by a widely occurring ketosynthase. Angew Chem Int Ed 52:4108–4112
Gokhale RS, Khosla C (2000) Role of linkers in communication between protein modules. Curr Opin Chem Biol 4:22–27
Grünewald J, Marahiel MA (2006) Chemoenzymatic and template-directed synthesis of bioactive macrocyclic peptides. Microbiol Mol Biol Rev 70:121–146
Gulick AM (2009) Conformational dynamics in the Acyl-CoA synthetases, adenylation domains of non-ribosomal peptide synthetases, and firefly luciferase. ACS Chem Biol 4:811–827
Hahn M, Stachelhaus T (2004) Selective interaction between nonribosomal peptide synthetases is facilitated by short communication-mediating domains. Proc Natl Acad Sci 101:15585–15590
Hahn M, Stachelhaus T (2006) Harnessing the potential of communication-mediating domains for the biocombinatorial synthesis of nonribosomal peptides. Proc Natl Acad Sci 103:275–280
Hang HC, Linder ME (2011) Exploring protein lipidation with chemical biology. Chem Rev 111:6341–6358
Hang HC, Geutjes E-J, Grotenbreg G, Pollington AM, Bijlmakers MJ, Ploegh HL (2007) Chemical probes for the rapid detection of fatty-acylated proteins in mammalian cells. J Am Chem Soc 129:2744–2745
Hernandez-Lucas I, Calva E (2012) The coming of age of the LeuO regulator. Mol Microbiol 85:1026–1028
Hu K, Li J, Li B, Webster JM, Chen G (2006) A novel antimicrobial epoxide isolated from larval Galleria mellonella infected by the nematode symbiont, *Photorhabdus luminescens* (Enterobacteriaceae). Bioorg Med Chem 14:4677–4681
Ikeda H, Ishikawa J, Hanamoto A, Shinose M, Kikuchi H, Shiba T, Sakaki Y, Hattori M, Ōmura S (2003) Complete genome sequence and comparative analysis of the industrial microorganism streptomyces avermitilis. Nat Biotechnol 21:526–531

Jewett JC, Bertozzi CR (2010) Cu-free click cycloaddition reactions in chemical biology. Chem Soc Rev 39:1272–1279

Ji D, Yi Y, Kang G-H, Choi Y-H, Kim P, Baek N-I, Kim Y (2004) Identification of an antibacterial compound, benzylideneacetone, from Xenorhabdus nematophilaagainst major plant-pathogenic bacteria. FEMS Microbiol Lett 239:241–248

Joyce SA, Brachmann AO, Glazer I, Lango L, Schwär G, Clarke DJ, Bode HB (2008) Bacterial biosynthesis of a multipotent stilbene. Angew Chem Int Ed 47:1942–1945

Kolb HC, Finn MG, Sharpless KB (2001) Click-Chemie: diverse chemische Funktionalität mit einer Handvoll guter Reaktionen. Angew Chem 113:2056–2075

Kontnik R, Crawford JM, Clardy J (2010) Exploiting a global regulator for small molecule discovery in *Photorhabdus luminescens*. ACS Chem Biol 5:659–665

Kresovic D, Schempp F, Cheikh-Ali Z, Bode HB (2015) A novel and widespread class of ketosynthase is responsible for the head-to-head condensation of two acyl moieties in bacterial pyrone biosynthesis. Beilstein J Org Chem 11:1412–1417

Lautru S, Challis GL (2004) Microbiology 150(Pt 6):1629–1636. Review. PMID: 15184549

Li J, Chen G, Wu H, Webster JM (1995) Identification of two pigments and a hydroxystilbene antibiotic from *Photorhabdus luminescens*. Appl Environ Microbiol 61:4329–4333

Ling LL, Schneider T, Peoples AJ, Spoering AL, Engels I, Conlon BP, Mueller A, Schäberle TF, Hughes DE, Epstein S et al (2015) A new antibiotic kills pathogens without detectable resistance. Nature 517:455–459

Marahiel MA (2016) A structural model for multimodular NRPS assembly lines. Nat Prod Rep 33:136–140

Mootz HD, Marahiel MA (1997) The tyrocidine biosynthesis operon of Bacillus brevis: complete nucleotide sequence and biochemical characterization of functional internal adenylation domains. J Bacteriol 179:6843–6850

Mootz HD, Schwarzer D, Marahiel MA (2002) Ways of assembling complex natural products on modular nonribosomal peptide synthetases. ChemBioChem 3:490–504

Newman DJ, Cragg GM (2012) Natural products as sources of new drugs over the 30 years from 1981 to 2010. J Nat Prod 75:311–335

Nollmann FI, Dauth C, Mulley G, Kegler C, Kaiser M, Waterfield NR, Bode HB (2015a) Insect-specific production of new GameXPeptides in *Photorhabdus luminescens* TTO1, widespread natural products in entomopathogenic bacteria. ChemBioChem 16:205–208

Nollmann FI, Heinrich AK, Brachmann AO, Morisseau C, Mukherjee K, Casanova-Torres ÁM, Strobl F, Kleinhans D, Kinski S, Schultz K et al (2015b) A *Photorhabdus* natural product inhibits insect juvenile hormone epoxide hydrolase. ChemBioChem 16:766–771

Omura S, Ikeda H, Ishikawa J, Hanamoto A, Takahashi C, Shinose M, Takahashi Y, Horikawa H, Nakazawa H, Osonoe T et al (2001) Genome sequence of an industrial microorganism streptomyces avermitilis: deducing the ability of producing secondary metabolites. Proc Natl Acad Sci 98:12215–12220

Pérez AJ, Wesche F, Adihou H, Bode HB (2016) Solid-phase enrichment and analysis of azide-labeled natural products: fishing downstream of biochemical pathways. Chemistry 22:639–645

Scharf DH, Brakhage AA (2013) Engineering fungal secondary metabolism: a roadmap to novel compounds. J Biotechnol 163:179–183

Schimming O, Fleischhacker F, Nollmann FI, Bode HB (2014) Yeast homologous recombination cloning leading to the novel peptides ambactin and xenolindicin. ChemBioChem 15:1290–1294

Schmitt EK, Hoepfner D, Krastel P (2015) Natural products as probes in pharmaceutical research. J Ind Microbiol Biotechnol 43:249–260

Schöner TA, Kresovic D, Bode HB (2015) Biosynthesis and function of bacterial dialkylresorcinol compounds. Appl Microbiol Biotechnol 1–6

Seo S, Lee S, Hong Y, Kim Y (2012) Phospholipase A2 inhibitors synthesized by two entomopathogenic bacteria, xenorhabdus nematophila and *Photorhabdus temperata* subsp. temperata. Appl Environ Microbiol 78:3816–3823

Sharon G, Garg N, Debelius J, Knight R, Dorrestein PC, Mazmanian SK (2014) Specialized metabolites from the microbiome in health and disease. Cell Metab 20:719–730

Sieber SA, Marahiel MA (2005) Molecular mechanisms underlying nonribosomal peptide synthesis: approaches to new antibiotics. Chem Rev 105:715–738

Stein ML, Beck P, Kaiser M, Dudler R (2012) One-shot NMR analysis of microbial secretions identifies highly potent proteasome inhibitor

Sucipto H, Wenzel SC, Müller R (2013) Exploring chemical diversity of α-Pyrone antibiotics: molecular basis of myxopyronin biosynthesis. ChemBioChem 14:1581–1589

Theodore CM, King JB, You J, Cichewicz RH (2012) Production of cytotoxic glidobactins/luminmycins by *Photorhabdus asymbioticain* liquid media and live crickets. J Nat Prod 75:2007–2011

Tobias NJ, Heinrich AK, Eresmann H, Wright PR, Neubacher N, Backofen R, Bode HB (2016a) Environ Microbiol. Aug 23. doi:10.1111/1462-2920.13502

Tobias NJ, Mishra B, Gupta DK, Sharma R, Thines M, Stinear TP, Bode HB (2016b) Genome comparisons provide insights into the role of secondary metabolites in the pathogenic phase of the *Photorhabdus* life cycle. BMC Genomics

Ullah I, Khan A, Ali L, Khan A, Waqas M, Lee I-J, Shin J-H (2014) An Insecticidal compound produced by an insect-pathogenic bacterium suppresses host defenses through phenoloxidase inhibition. Molecules 19:20913–20928

Ullah I, Khan AL, Ali L, Khan AR, Waqas M, Hussain J, Lee I-J, Shin J-H (2015) Benzaldehyde as an insecticidal, antimicrobial, and antioxidant compound produced by *Photorhabdus temperata* M1021. J Microbiol 53:127–133

Vogel J, Luisi BF (2011) Hfq and its constellation of RNA. Nat Rev Micro 9:578–589

The Regulation of Secondary Metabolism in *Photorhabdus*

David J. Clarke

Abstract A general definition of secondary metabolism is that it consists of the metabolic pathways and the products of metabolism that are not absolutely required for the survival of the organism. Using this definition, it is now well established that *Photorhabdus* elaborate an extensive secondary metabolism during the post-exponential phase of bacterial growth. This secondary metabolism includes, but is not limited to, the production of light, a stilbene antibiotic and an anthraquinone pigment. In this chapter, the role of secondary metabolism during the life cycle of *Photorhabdus* will be discussed. Recent work has shown that secondary metabolism in *Photorhabdus* is required for the mutualistic association between the bacteria and its nematode partner, in particular bacterial secondary metabolism is required to support normal nematode growth and development. An isogenic population of *Photorhabdus* is phenotypically heterogenous and this facilitates functional partitioning within the population. The relationship between secondary metabolism and the various phenotypic and phase variants that exist in populations of *Photorhabdus* will also be discussed. Finally, this chapter will also describe the various regulatory nodes that have been identified as being part of the complex regulatory network that is used to control the temporal expression of secondary metabolism in *Photorhabdus*.

Contents

D.J. Clarke (✉)
School of Microbiology and APC Microbiome Institute,
University College Cork, Cork, Ireland
e-mail: david.clarke@ucc.ie

Current Topics in Microbiology and Immunology (2017) 402:81–102
DOI 10.1007/82_2016_21

Published Online: 26 July 2016

1 Introduction

Secondary metabolism can be defined as consisting of the metabolic pathways and the products of metabolism that are not absolutely required for the survival of the organism. Many bacteria, e.g. *Streptomyces,* have important secondary metabolisms that are responsible for the production of a range of clinically important molecules, such as antibiotics and immunosuppressive drugs. Secondary metabolism is generally associated with the post-exponential or stationary phase of bacterial growth when nutrients are limited and metabolic flux is redirected away from the production of new bacterial cells (i.e. primary metabolism) to the production of secreted metabolites (i.e. secondary metabolism). However, there is no reason why secondary metabolites cannot also be produced during bacterial growth and it is now clear that secondary metabolism in bacteria is regulated by a complex network of global and pathway-specific regulators that link primary and secondary metabolism. The role of secondary metabolism (and the function of the resulting secondary metabolites) during the normal life of a bacterium is not fully understood but these molecules may act as signals to modulate the interaction between the producing bacterium and other organisms in the ecosystem (Romero et al. 2011).

Members of the *Photorhabdus* genus have an extensive secondary metabolism that includes the production of a wide range of different activities during the stationary phase of bacterial growth (Clarke 2008; Waterfield et al. 2009; Bode 2009). These activities include the genus defining characteristic of bioluminescence (i.e. light production) as well as the production of a number of small bioactive molecules such as a multipotent stilbene and a polyketide pigment called anthraquinone (AQ). The metabolic pathways responsible for the production of both stilbene and AQ have been described and will be discussed elsewhere in this book (see Chap. 20) (Brachmann et al. 2007; Joyce et al. 2008). Many secondary metabolites are produced through the action of enzymes called polyketide synthases (PKS) and non-ribosomal peptide synthases (NRPS) and recent work has identified a significant number of genetic loci encoding potential PKS, NRPS and PKS-NRPS chimera on the *Photorhabdus* genome (Bode 2009). Many of these loci are cryptic and therefore the genes are not expressed under normal laboratory conditions (Brachmann et al. 2012a). This suggests that the expression of these loci is tightly regulated in *Photorhabdus*. Nonetheless, a range of biochemical and genetic approaches have been used to identify and characterize a significant number of the molecules produced by the enzymes encoded at several of these loci (see Table 1 and references therein).

Table 1 Secondary metabolites produced by *Photorhabdus luminescens* TTO1

Metabolite	Genetic loci	Role	References
Stilbene	*stlA; stlB; stlCDE; bkdABC*	Antibiotic; IJ recovery signal; PPO inhibitor	Williams et al. (2005) and Joyce et al. (2008)
Pyrone[b]	*ppyS*	Quorum sensing	Brachmann et al. (2013)
Lumiquinone A	*plu2233*?	Unknown	Park and Crawford (2015)
Dihydrophenylalanine (dihydrostilbene)	*plu3042-3044*	Unknown	Crawford et al. (2011)
Anthraquinone	*antABCDEFGHI*	Redox active; possible mosquitocidal activity	Brachmann et al. (2007) and Ahn et al. (2013)
Glidobactin/Cepafungin I	*plu1877-1881*	Proteasome inhibitors	Bian et al. (2012), Theodore et al. (2012), Dudnik et al. (2013) and Stein et al. (2012)
Rhabduscin	*plu1760-1762; insA-insB*	PPO inhibitor	Crawford et al. (2012)
Phurealipids	*pliA*	Juvenile hormone epoxide hydrolase inhibitor	Nollmann et al. (2015)
Indigoidine[a]	*plu2180-2185-indC-indA*	Blue pigment	Brachmann et al. (2012a)
GameXpeptides/luminmides	*plu3263; plu3561-3567*	Unknown	Bode et al. (2012) and Fu et al. (2012)
Kollisin A[a]	*kol (plu2670)*	Unknown	Bode et al. (2015)

[a]Cryptic
[b]*P. asymbiotica* use dialkylresorcinol instead of pyrone as a QS signalling molecule (synthesized by *darABC* [same as *stlCDE* in *P. luminescens* TTO1)]

2 The *Photorhabdus* Life Cycle

Photorhabdus has a complex life cycle that involves alternating and temporally distinct pathogenic and mutualistic interactions with insect and nematode hosts, respectively (Goodrich-Blair and Clarke 2007; Waterfield et al. 2009; Clarke 2014). The bacteria are normally found in the gut lumen of the infective juvenile (IJ) of the soil-dwelling nematode *Heterorhabditis*. The IJ is a specialized developmental stage of the nematode, similar to the dauer juvenile of the well-studied free living nematode *Caenorhabditis elegans*. The IJ actively seeks out and infects suitable insect larvae before releasing the bacteria into the blood (i.e. hemolymph) of the infected insect. The regurgitation of the bacteria is a response to a small, unidentified signal present in the insect hemolymph and regurgitation signals the start of a developmental pathway, called IJ recovery, that results in the formation of a self-fertile adult hermaphrodite from every IJ (Ciche and Ensign 2003). The bacteria replicate in the insect and, whilst doing so, convert the internal organs and tissues of the insect into bacterial biomass. The insects generally die within 48–72 h post-infection. Bacterial growth in the insect is exponential and there is a very strong positive correlation between bacterial growth rate and the time taken to kill an insect (generally represented as an LT_{50}) suggesting that virulence is associated with bacterial growth and, therefore, primary metabolism (Clarke and Dowds 1995; Watson et al. 2005). The bacteria replicate to a high cell density within the insect and, as in any batch system, the bacterial population eventually enter the post-exponential (or stationary) phase of growth. This generally coincides with the completion of IJ recovery. At this time, the adult hermaphrodite nematodes are present in the insect cadaver with a high density of *Photorhabdus* bacteria cells (Hu and Webster 2000). The hermaphrodite will externally lay about 300 eggs that hatch within the bacterial biomass present in the insect cadaver and develop through several juvenile moults (J1–J4) to adulthood. Normally, there are 2–3 generations of nematode reproduction in the insect and during this time the nematodes feed on the bacterial biomass. The adult hermaphrodite does not lay all of her eggs and the retained eggs will hatch within the body cavity of the maternal nematode in a process called *endotokia matricida* (Ciche et al. 2008). The internally hatched nematodes develop exclusively into the IJ stage and, during development in the hermaphrodite, each IJ is colonized by *Photorhabdus*. The colonization of the IJs (i.e. symbiont transmission) is controlled by a phase variable invertible DNA switch in *Photorhabdus* called the *madswitch* (Somvanshi et al. 2010, 2012). The *madswitch* is a binary switch that can exist in one of 2 orientations, i.e. ON or OFF. When the *madswitch* is ON, the bacteria produce surface-localized fimbriae, called the Mad fimbriae, that facilitate adherence to the intestinal wall of the maternal nematode and subsequent colonization of the IJ. Each IJ is normally initially colonized by 1 bacterial cell and this bacterial cell multiplies in the gut lumen of the nematode such that mature IJs contain, on average, approximately 100 bacterial cells (Somvanshi et al. 2010). The complete life cycle normally takes 10–20 days and >100,000 IJs will emerge from an insect originally infected with a single IJ highlighting the remarkable efficiency of this symbiosis.

3 Secondary Metabolism, Pathogenicity and Mutualism

The mutualistic interaction between *Photorhabdus* and *Heterorhabditis* coincides with the post-exponential phase of bacterial growth in the insect (Hu and Webster 2000). Therefore, mutualism also coincides with the activation of *Photorhabdus* secondary metabolism, thus temporally connecting secondary metabolism and mutualism (Joyce et al. 2011). What is the role of secondary metabolism in mutualism? Perhaps the most significant evidence for a role for secondary metabolism in mutualism was extrapolated from early observations of phenotypic variation in *Photorhabdus* (Akhurst 1980; Boemare and Akhurst 1988). *Photorhabdus* can exist as 2 stable phenotypic variants when cultured on agar plates, i.e. the primary and the secondary variants. The primary variant has a normal secondary metabolism (including the production of AQ, stilbene and light) and is both pathogenic to insects and capable of supporting nematode growth and development. The secondary variant, on the other hand, does not produce AQ, stilbene or light and, whilst this variant is pathogenic, it is unable to support nematode growth and development. Therefore, it has been proposed that secondary metabolism (as characterized in the primary variant) is important for the mutualistic association with the nematode (Ffrench-Constant et al. 2003). However, only one secondary metabolite has been shown to have a direct role in mutualism, i.e. the stilbene antibiotic. During in vitro symbiosis assays (where the *Photorhabdus* bacteria are inoculated onto lipid agar plates and incubated for 3–4 days before the biomass is seeded with IJ stage nematodes), mutants unable to produce stilbene do not support IJ recovery to adult hermaphrodites (Williams et al. 2005; Joyce et al. 2008). This suggests that the stilbene is an inter-kingdom signalling molecule that is produced by the bacteria and is perceived by the IJ nematode and signals initiation of the recovery programme. However, during growth in the insect, the role of the stilbene may be to act as a food signal that stimulates the recovery of IJs that develop during the normal reproductive cycle of the nematodes (Strauch and Ehlers 1998). This would function to link nematode reproduction and development with the availability of food, i.e. *Photorhabdus* biomass. The primary variant of *Photorhabdus* also produces 2 crystalline inclusion proteins that are rich in essential amino acids during post-exponential growth leading to the suggestion that these proteins provide important nutrition to the nematode (Bowen and Ensign 2001; You et al. 2006). However, in vitro studies suggest that, with the exception of stilbene, the other activities associated with secondary metabolism in *Photorhabdus* may not be required for nematode growth and development (Han and Ehlers 2001; Bager et al. 2016). The bottleneck for nematode growth and development when cultured on lipid agar plates inoculated with *Photorhabdus* appears to be IJ recovery, a process that during in vitro symbiosis assays is almost entirely dependent on the production of stilbene. Therefore, mutants that do not express the normal secondary metabolism exhibit a decrease in nematode growth and development due to the reduced production of stilbene and lower levels of IJ recovery (Bager et al. 2016). Nonetheless, it is possible that many activities associated with secondary metabolism in *Photorhabdus* may have non-nutrition-based roles in the interaction with the nematode during growth in the insect, e.g. the production of signals that protect the insect cadaver niche (Fenton et al. 2011; Jones et al. 2016).

Whilst secondary metabolism is normally associated with the post-exponential phase of bacterial growth, there is evidence that some secondary metabolites are produced during exponential phase (i.e. during bacterial growth). The growth rate of *Photorhabdus* has been correlated with virulence suggesting that any secondary metabolite produced during this phase of growth may have a role in virulence. The expression of the *cpm* genes encoding the proteins required for the production of the carbapenem antibiotic by *Photorhabdus* is maximal during exponential growth and decreases as the cells enter post-exponential growth (Derzelle et al. 2002). Whilst unlikely to be a virulence factor *per se*, the production of this antibiotic would be expected to increase the fitness of *Photorhabdus* by killing off the indigenous insect microbiota. In addition, biochemical and genetic analyses have identified a very potent antagonist of prophenoloxidase (PPO), called rhabduscin, that is produced by *Photorhabdus*. PPO is a key enzyme in the insect innate immune system that catalyses the polymerization of melanin resulting in the death and/or nodulation/encapsulation of invading organisms (Cerenius and Söderhäll 2004; Kanost et al. 2004; Cerenius et al. 2008). A deletion of the genes required for rhabduscin biosynthesis in the closely related entomopathogenic bacterium *Xenorhabdus* did result in a decrease in virulence suggesting that this compound may be a virulence factor. The rhabduscin was shown to be physically associated with the outer surface of the bacteria resulting in the formation of a "defensive perimeter" (Crawford et al. 2012). Rhabduscin has been detected in *P. luminescens* TTO1 culture supernatants although the kinetics of rhabduscin production has not been measured (Crawford et al. 2010).

4 Phenotypic Variation, Phase Variation and Secondary Metabolism

As previously mentioned (nearly), every IJ is initially colonized by a single bacterium and, therefore, an infected insect is expected to contain a clonal population of *Photorhabdus*. However, there is a significant degree of functional specialization within this clonal population. It has already been mentioned that *Photorhabdus* can exist in 2 relatively stable phenotypic variants when cultured on agar plates, i.e. the primary and secondary variants. It is now well established that many bacteria exhibit phenotypic variation (i.e. bistability) as a risk-limiting adaptation that increases the adaptability of a population of isogenic bacteria to changing environments (Dubnau and Losick 2006; Veening et al. 2008; Ackermann 2015). This survival strategy when displayed by isogenic populations is called bet-hedging. The rate of formation of the secondary variant has been shown to increase as the population of *Photorhabdus* gets older (e.g. over timescales that are similar to the time required for nematode growth and development) and the kinetics of phenotypic variation is controlled by the AstSR two-component pathway (2CP) (Derzelle et al. 2004). Mutants in the *astR* gene (encoding the predicted response regulator) exhibit increased and promiscuous (i.e. early) formation of the secondary variant, and proteomic analysis of this mutant suggests that this phenotype may be linked to defects in redox homoeostasis. Therefore, the AstSR 2CP may control phenotypic variation is response to nutrient limitation and redox stress (Derzelle et al. 2004). Interestingly, the AstSR 2CP does not regulate secondary metabolism directly (i.e. the *astR* mutant produces the same level of light, stilbene and AQ as the wild type) but this pathway can obviously control (through phenotypic variation) the level of secondary metabolism within a population of *Photorhabdus*.

The vast majority of *Photorhabdus* inside an insect will not have the opportunity to colonize the IJ and these bacteria will remain in the insect cadaver after the nematodes have left (see Fig. 1). As secondary metabolism is associated with life with the nematode, the switch from primary variant to secondary variant (with the associated switching off of secondary metabolism) may be an adaptation for life in the environment without the nematode. This hypothesis is supported by comparisons of the metabolism and proteomes of primary and secondary variants that suggest that the secondary variant is metabolically more robust than the primary variant during growth in vitro and in vivo (Smigielski et al. 1994; Turlin et al. 2006) (Susan Joyce, unpublished data).

In addition to phenotypic variation, *Photorhabdus* has also been shown to exhibit phase variation, mediated by invertible DNA elements, at 2 different genetic loci, i.e. the *mrf* operon (encoding mannose-resistant fimbriae) and the *mad* operon (encoding the Mad fimbriae) (Meslet-Cladiere et al. 2004; Somvanshi et al. 2010). As already mentioned, the production of the Mad fimbriae is controlled by the orientation of a DNA element called the *madswitch*. However, the *madswitch* also controls the production of a number of additional phenotypes, including secondary metabolism (Somvanshi et al. 2012). *Photorhabdus* with the *madswitch* in the OFF orientation are phenotypically equivalent to the primary variant (hereafter called the P-form). The P-form has a normal secondary metabolism and is both pathogenic to

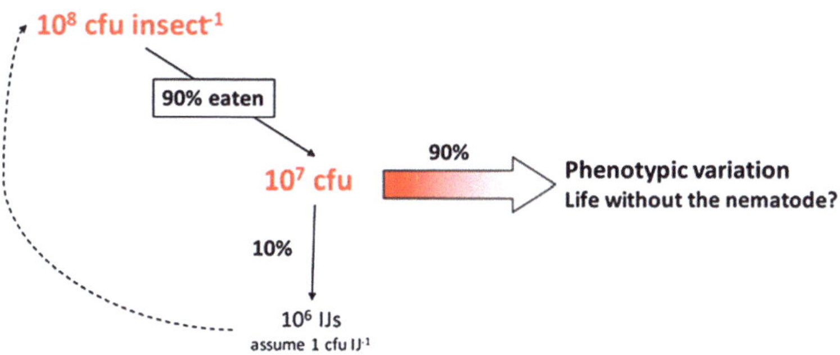

Fig. 1 Secondary variant is adapted for life without the nematode. *Photorhabdus* achieve a final cell density of approximately 10^8 bacterial cells [colony forming units (cfu)] per insect. It is not unreasonable to assume that 90 % of the bacterial biomass will be eaten by the nematodes that are developing within the insect cadaver. This scenario would suggest that 10^7 cfu could remain in the insect at the end of the vegetative life cycle of the nematode. Assuming that each insect supports the development of 10^6 IJs and each IJ is colonized by a single *Photorhabdus* cell then a very conservative estimation is that 9×10^6 cfu remain in the insect cadaver after the nematode has left. A simplistic model would suggest that these bacteria are converted to the secondary variation in order to adapt to life without the nematode. However, it is more likely that secondary variants will accumulate during growth of the bacteria in the insect (a phenomenon regulated by HexA and AstSR) and the proposed model suggests that the system is buffered so that phenotypic variation is tolerated, up to a point, without deleterious consequences to nematode growth

insects and able to support nematode growth and development. However, the P-form is not able to colonize the IJ as it is not producing the Mad fimbriae. On the other hand, *Photorhabdus* with the *madswitch* in the ON orientation (i.e. M-form) are able to colonize the IJ. However, the M-form bears some phenotypic resemblance to the secondary variant, i.e. the M-form does not produce light, the stilbene antibiotic or the AQ pigment. Therefore, a mixture of P-form and M-form bacteria is essential for the complete mutualistic association with the nematode. At some point after the colonization of the adult hermaphrodite, it has been shown that the M-form reverts back to the P-form (Somvanshi et al. 2010). This ensures that the IJ is colonized by the P-form so that the nematodes will have the required nutrition following infection of a new insect host. In addition, there is some evidence to suggest that the formation of the M-form, like the secondary variant, is linked to nutrient limitation and/or oxidative stress (Somvanshi et al. 2012; Blackburn et al. 2016). However, the M-form is not virulent to insects and M-form bacteria form small colonies [i.e. small colony variants (SCVs)] on agar plates compared to the P-form and secondary variant (see Fig. 2). Therefore, the M-form and the

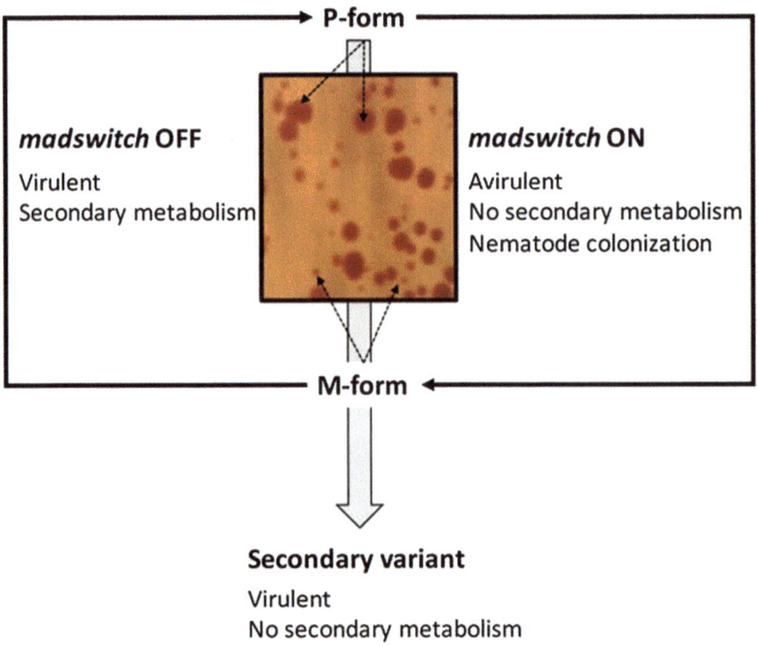

Fig. 2 Phenotypic and phase variation in *P. luminescens* TTO1. A population of *P. luminescens* exists as a mixture of P-form and M-form cells determined by the orientation of a binary DNA switch, the *madswitch*. When plated on agar, the P-form cells form large colonies whilst the M-form cells form small colonies (as indicated). The P-form is virulent and produces a secondary metabolism (i.e. supports nematode growth and development) whilst the M-form is adapted for colonization of the nematode and, therefore, it is avirulent and does not have a secondary metabolism. Photorhabdus also form a secondary variant but the origins of the secondary variant are unclear; i.e. is it derived from the P-form or M-form? Indeed, the secondary variant shares phenotypes with both the M-form and the P-form but these variants are distinct

secondary variant are not equivalent and the regulatory networks controlling secondary metabolism during phenotypic variation and phase variation at the *mad* locus would appear to be different.

Nonetheless, the complex phenomena of phenotypic and phase variation ensure that there is functional heterogeneity within a clonal population of *Photorhabdus*, such as exists in the insect cadaver. This enables the bacteria to carry out several distinct functions that are essential for mutualism with the nematode and the survival of the bacteria, i.e. the ability to express secondary metabolism and support nematode growth and development, the ability to colonize the next generation of IJs (i.e. transmission) and the ability to persist in the environment in the absence of the nematode.

5 The Regulation of Secondary Metabolism

It is clear that secondary metabolism is differentially expressed in the different phenotypic and phase variants of *Photorhabdus* that have been characterized to date. Studies aimed at describing the structure of the regulatory network(s) controlling secondary metabolism have been undertaken and key players (i.e. regulatory nodes) have been identified. Most of these studies have been undertaken in the primary variant where efforts have been made to understand the regulatory link between secondary metabolism and mutualism and important global and pathway-specific regulators have been identified. Some of these regulators have also been implicated in the regulation of secondary metabolism in the secondary variant and M-form. Finally, it is worth highlighting that the vast majority of these studies have been done using the de facto type strain of *Photorhabdus*, *P. luminescens* TTO1.

5.1 Global Regulators

5.1.1 HexA

A key regulator of secondary metabolism in *Photorhabdus* is a LysR-type transcriptional regulator called HexA. This regulator was initially identified as a repressor of secondary metabolism in the secondary variant. Interruption of the *hexA* gene in the secondary variant of *P. temperata* K122 restored the expression of secondary metabolism (and the ability to support nematode growth and development) to this strain (Joyce and Clarke 2003). However, HexA also appears to have a regulatory role in the primary variant where it is required for the correct temporal regulation of secondary metabolism. Therefore, a *hexA* deletion mutant in the primary variant of *P. luminescens* TTO1 produces increased levels of light, stilbene and derivatives of the stilbene molecule throughout growth and not just in the

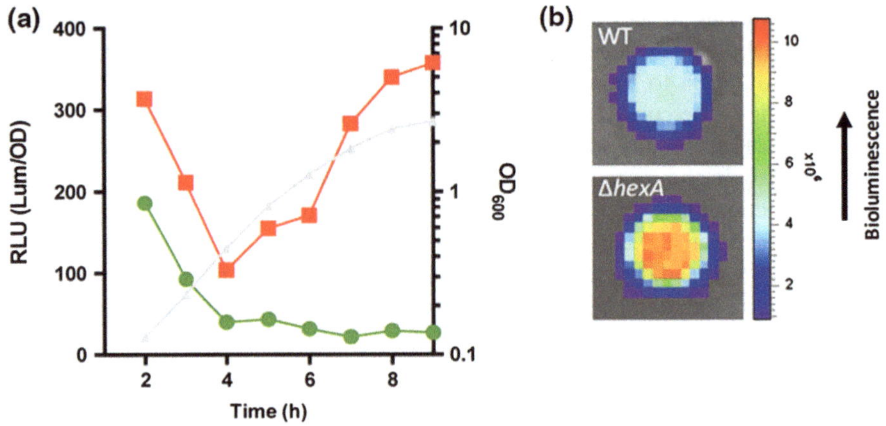

Fig. 3 HexA represses bioluminescence during growth of the *P. luminescens* TTO1 primary variant. **a** Light production (reported as Relative Light Units (RLU)) from *P. luminescens* primary variant (*green circles*) and Δ*hexA* mutant (*red squares*) was measured during growth in LB broth at 28 °C with shaking. There were no differences in the growth of either strain under these conditions (*grey triangles*). **b** The light produced by colonies of TTO1 primary variant and the Δ*hexA* mutant was measured after growth on LB agar for 72 h. Light production from the Δ*hexA* mutant is clearly increased compared to the primary variant control. Bioluminescence was measured using an In Vitro Imaging System (IVIS-100, Xenogen Alameda, CA, USA)

post-exponential phase (Kontnik et al. 2010) (see Fig. 3). Interestingly, HexA repressed AQ production in *P. temperata* but not in *P. luminescens* suggesting species-specific regulatory targets (Kontnik et al. 2010). HexA may also have a role in repressing secondary metabolism in the M-form as transcriptome analysis has indicated that the expression of *hexA* is increased by 4.4-fold in a strain that is genetically locked into the M-form (Somvanshi et al. 2012). Therefore, HexA appears to be an important repressor of secondary metabolism in *Photorhabdus*. However, the regulatory network(s) controlled by HexA in the primary and secondary variants (and potentially the M-form) are not well understood.

5.1.2 The BarA-UvrY Two-Component Pathway

The BarA-UvrY 2CP activates the transcription of genes involved in secondary metabolism in *Photorabdus*. A deletion mutant in *uvrY* (encoding the response regulator) in *P. luminescens* exhibited much lower levels of expression of the genes required for bioluminescence, stilbene and AQ biosynthesis (Krin et al. 2008). However, many of the transcriptional changes observed in the *uvrY* mutant strain were not evident at either the protein or phenotypic level suggesting that the expression of these genes may also be significantly regulated at the post-transcriptional level (Krin et al. 2008; Kontnik et al. 2010; Lango-Scholey et al. 2013). Indeed, to this end, the BarA-UvrY 2CP in other bacteria has been shown to work in association with small regulatory RNA molecules. A well-studied target of the BarA-UvrY 2CP is the CsrA-*csrB* regulatory system (Pernestig et al. 2003; Weilbacher et al. 2003). In this system, CsrA is a RNA-binding protein that represses translation by binding to the ribosome-binding site of the target mRNA (Liu and Romeo 1997). The *csrB* RNA carries multiple target sites from CsrA and, when expressed, titrates CsrA away from target mRNAs, thus increasing the translation efficiency of these transcripts (Romeo 1998; Babitzke and Romeo 2007). In *P. luminescens* TTO1, the CsrA-*csrB* regulatory system was shown to play a role in the regulation of some, but not all, of the identified UvrY targets (Krin et al. 2008). The BarA-UvrY 2CP is also known as GacS-GacA in other bacteria such as *Pseudomonas* where the GacS-GacA 2CP has also been shown to have an important role in the regulation of secondary metabolism (Lapouge et al. 2007). Similar to what has been observed in *Photorhabdus*, mutations in the GacS-GacA 2CP (a pathway that works together with the Rsm RNA regulatory system) block secondary metabolism in several different species of pseudomonads (Heeb and Haas 2001). Interestingly, recent work has also shown that *gacS* mutations in the rhizobacterium *Pseudomonas aeruginosa* M18 generate phenotypic diversification resulting in the formation of SCVs both in vivo and in vitro (Davies et al. 2007; Nelson et al. 2010). Moreover, the GacS-GacA system was shown to have an interesting role in the association between *Ps. fluorescens* and the bacterial-feeding, social amoeba *Dictyostelium discoideum*. In this system, many *D. discoideum* have been observed to engage in bacterial husbandry by carrying and disseminating the bacteria during the spore stage, thus ensuring that there is food available for the amoeba following germination of the spores. In initial studies, *D. discoideum* was shown to carry 2 strains of *Ps. fluorescens* although only 1 strain served as a food source for the amoeba (Brock et al. 2011). Recently, it was shown that the second strain produced secondary metabolites, chromene and pyrrolnitrin that were beneficial to the amoeba by enhancing appropriate spore formation. Intriguingly the secondary metabolite-producing strain could be converted into the food source by a loss-of-function mutation in the *gacA* gene (Stallforth et al. 2013). Therefore, the *gacA* mutation generates functional heterogeneity within the population of *Ps. fluorescens*. *Photorhabdus* has similar roles during the mutualistic association with the nematode (i.e. it is a food source for the nematode and it produces molecules that control nematode development). Although there is no evidence to suggest that non-functional alleles of *uvrY* (the homologue of *gacA*) emerge and co-exist within insect cadavers it is possible that the activity of the BarA-UvrY 2CP is modulated (e.g. post-transcriptionally) to induce functional heterogeneity within the bacterial population (Camacho et al. 2015). Nonetheless, it is clear that the homologous BarA-UvrY and GacS-GacA 2CPs play a key role in the regulation of functional heterogeneity and secondary metabolism in different bacteria–host interactions.

5.1.3 Magic Spot-(P)ppGpp and Nutrient Limitation

A recent study in *Photorhabdus* showed that the expression of *stlA*, encoding phenylalanine ammonium-lyase (PAL), was regulated by the availability of nutrients, in particular amino acids (Lango-Scholey et al. 2013). PAL is the first enzyme in the synthesis of the stilbene antibiotic and this enzyme converts the amino acid phenylalanine into cinnamic acid (CA) further highlighting the connection between amino acid metabolism and secondary metabolism (Williams et al. 2005). It is well established throughout the gammaproteobacteria that nutrient limitation results in the production of a small signalling molecule (or alarmone) called (p)ppGpp (Gaca et al. 2015a; Hauryliuk et al. 2015). The production of (p)ppGpp generally results in the halting of protein synthesis and DNA replication and, through an interaction with RNA polymerase, the biasing of transcription towards genes required for adaptation to the stress. The level of (p)ppGpp in the bacterial cell is controlled by the activity of two proteins, RelA and SpoT. Recent work has shown that RelA monitors the aminoacylated-status of the tRNA present in the A site of the ribosome (Brown et al. 2016). RelA is activated when uncharged tRNA molecules enter the A-site (indicating amino acid limitation) and this results in the synthesis of (p)ppGpp. RelA appears to respond primarily to amino acid starvation whilst SpoT integrates various other signals into (p)ppGpp metabolism. Moreover, the hydrolysis activity of SpoT is essential for maintaining appropriate levels of (p)ppGpp during growth (Hauryliuk et al. 2015).

The accumulation of (p)ppGpp has been shown to result in significant transcriptional changes in many bacteria (Hesketh et al. 2007; Traxler et al. 2010; Vercruysse et al. 2011; Bowden et al. 2013; Gaca et al. 2015b). In *E. coli,* structural and genetic analyses suggest that the small alarmone interacts directly with RNA polymerase although the exact molecular details of how this interaction affects transcription are not fully understood (Zuo et al. 2013; Ross et al. 2013). The synthesis of (p)ppGpp is often linked to nutrient limitation and growth arrest and this is likely to coincide with a change from primary to secondary metabolism. In a recent study, the accumulation of (p)ppGpp was shown to be required for secondary metabolism and mutualism in *Photorhabdus* (Bager et al. 2016). Therefore, a *relA spoT* double mutant was constructed and shown to be completely unable to synthesis (p)ppGpp. The *relA spoT* mutant did not produce light, stilbene or AQ pigment and was also unable to support nematode growth and development in vitro. This suggests, perhaps paradoxically (given the role for *Photorhabdus* in nematodes nutrition), that nutrient limitation in *Photorhabdus* is important for mutualism. In contrast, the *relA spoT* mutant was as virulent to insects as the wild-type strain indicating the alarmone production was not required for pathogenicity. However, the *relA spoT* mutant was rapidly out-competed by the wild-type strain during prolonged incubation in the insect again highlighting the important role for (p)ppGpp during nutrient limitation (Bager et al. 2016). The (p)ppGpp alarmone has also been shown to be required for the production of secondary metabolites such as the blue polyketide pigment, actinorhodin and the calcium-dependent antibiotic (CDA) in the model actinomycete, *Streptomyces coelicolor* (Hesketh et al. 2007). Moreover, a *relA spoT* mutant in *Ps. fluorescens* CHAO had decreased antibiotic activity and biocontrol activity suggesting a role for (p)ppGpp in the regulation of secondary metabolism in this bacterium (Takeuchi et al. 2012). Indeed, it seems that the level of (p)ppGpp in *Ps. fluorescens* CHAO can be regulated by the GacA-GacS 2CP system highlighting the complexity and interconnectedness of the networks involved in controlling secondary metabolism (Takeuchi et al. 2012).

5.1.4 Metabolism and the Metabolic Switch

Primary metabolism is generally associated with the central metabolic pathways (i.e. glycolysis, the TCA cycle) that produce energy and metabolic precursors to facilitate cell growth and reproduction. Secondary metabolism, on the other hand, can be considered as an alternative route for carbon and energy flux that leads to the production of carbon-rich bioactive compounds (rather than the production of bacterial biomass). Therefore, it is entirely appropriate that primary and secondary metabolism should be tightly linked, in terms of both regulatory and metabolic connections. Deletion of either the *mdh* or *fumC* gene and encoding key enzymes in the TCA cycle (malate dehydrogenase and fumarase, respectively) were shown to block the production of light, stilbene antibiotic and AQ pigment in *P. luminescens* TTO1 (Lango and Clarke 2010). As expected both the *mdh* and *fumC* mutants were also unable to support nematode growth and development in vitro or in vivo and both mutants were unaffected in virulence. This led to the suggestion that the transition from pathogenicity (i.e. *Photorhabdus* growth) to mutualism (i.e. *Photorhabdus* secondary metabolism) was controlled by a metabolic switch that involved the TCA cycle. Interestingly, a similar block in the TCA cycle in *Ps. fluorescens* CHAO (specifically a mutation in the *fumA* gene encoding fumarase) also blocked secondary metabolism in this bacterium where it was shown that an imbalance in the TCA cycle (leading to an accumulation of certain TCA cycle intermediates) prevented the appropriate activation of the GacS-GacA 2CP (Takeuchi et al. 2009). Similarly, a mutation in the *acnB* gene (encoding aconitase, a key TCA cycle enzyme) of the bioluminescent bacterium *Vibrio fischeri* resulted in increased bioluminescence and this was shown to be dependent on the Gac/Rsm regulatory system (Septer et al. 2015). Therefore, the BarA-UvrY (GacS-GacA) 2CP does appear to constitute a clear regulatory link between primary and secondary metabolism in many different bacteria.

L-proline is an abundant free amino acid in insect haemolymph and increasing concentrations of L-proline upregulate the production of the stilbene antibiotic and AQ pigment in *P. luminescens* TTO1 (Crawford et al. 2010). This phenotype was shown to be dependent on proline uptake and there is some evidence that the role of proline in secondary metabolism may be regulatory as L-proline may interact with HexA (Kontnik et al. 2010). However, L-proline assimilation also contributes to the generation of proton motive force (PMF) and dissipating the PMF with either CCCP or valinomycin resulted in increased levels of stilbene antibiotic and the AQ pigment suggesting that the electrical component of the proton motive force is linked to secondary metabolism (Crawford et al. 2010). Therefore, the presence of L-proline may have a role in coordinating the link between the PMF and the secondary metabolism in *Photorhabdus*.

5.1.5 Quorum Sensing and AI-2

Quorum sensing is a phenomenon in bacteria that is characterized by the production of signal molecules that mediate cell-to-cell communication and coordinates gene expression (Miller and Bassler 2001; Ng and Bassler 2009). *Photorhabdus* do not communicate using canonical quorum-sensing signals such as acylhomoserine lactones (AHLs). However, many bacteria have also been shown to quorum sense using an alternative signal called AI-2 (Pereira et al. 2013). This signal is produced by the product of the *luxS* gene, S-ribosylhomocysteine lyase, that has an important metabolic function during the activated methyl cycle where LuxS converts S-D-ribosyl-L-homocysteine into L-homocysteine. A product of this reaction is AI-2 and this molecule is excreted from the cell where it accumulates in the extracellular environment. The production of AI-2 has been shown to induce transcriptional changes in some bacteria leading to the suggestion that AI-2 may function as a signalling molecule (Kendall et al. 2007; Jesudhasan et al. 2010; Hirano et al. 2012). *Photorhabdus*, as is the case with many bacteria, has AI-2 receptors that bind to the secreted molecule and, through a link with AI-2 uptake systems, actively transport AI-2 back into the cell (Krin et al. 2006). In *Photorhabdus,* AI-2 has been implicated in the production of the carbapenem antibiotic (produced by genes found in the *cpm* operon) where a *luxS* deletion mutant was shown to express higher levels of *cpm* mRNA throughout growth (Derzelle et al. 2002). Transcriptomic analysis also revealed differences in the expression of >100 genes in the *luxS* mutant when compared to wild-type *P. luminescens* TTO1 (Krin et al. 2006). Interestingly, the *luxS* mutant produced less luminescence than the wild-type and this was attributed to the increased production of polyamines in the mutant. Moreover, the *luxS* mutant was more sensitive to oxidative stress, formed reduced biofilms and was hyper-motile when compared to the wild-type (Krin et al. 2006). Interestingly, the BarA-UvrY 2CP in *Photorhabdus* is required for maximal production of AI-2 and some of the phenotypes reported in the *uvrY* mutant (e.g. reduced bioluminescence) could be rescued by the addition of exogenous AI-2 (Krin et al. 2008). Therefore, AI-2 has pleiotropic roles in *Photorhabdus* that are linked to secondary metabolism and stress resistance.

5.2 *Pathway-specific Regulators*

In contrast to work described here on the global regulators of secondary metabolism in *Photorhabdus,* very little work has been done to identify regulators of specific secondary metabolic pathways, i.e. pathway-specific regulators. This type of regulatory network topology, whereby master or global regulators work with pathway-specific regulators to ensure that the expression of secondary metabolites is appropriate, is common in bacteria.

5.2.1 AQ

The AQ pigment is produced by proteins encoded by the 9-gene *antA-I* locus (Brachmann et al. 2007). Genes at both ends of this locus (*plu4185* and *plu4195*) are predicted to encode transcriptional regulators although any role for these genes in the regulation of AQ production remains to be established. Another transcriptional regulator, HdfR, was shown to act as a repressor of *antA-I* expression and AQ production (Easom and Clarke 2012). The *hdfR* gene was originally identified during a screen for mutants unable to colonize the IJ nematode and the role, if any, of AQ during nematode colonization is unclear.

5.2.2 Stilbene

The stilbene molecule produced by all *Photorhabdus* strains is a multipotent molecule with confirmed roles in pathogenicity and mutualism. The biosynthetic pathway of this molecule has, for the most part, been elucidated and involves proteins encoded by genes located in at least 4 different genetic loci (Joyce et al. 2008). Therefore, it is possible that the different loci may be regulated independently from each other and flux towards stilbene synthesis may involve both genetic and allosteric regulation. This is more likely when one considers that some of the loci, in addition to having a role in stilbene biosynthesis, are also required for the biosynthesis of fatty acids required during growth, i.e. the *bkdABC* operon encoding a branched-chain amino acid dehydrogenase (Brachmann et al. 2012b). As previously mentioned, the first committed step in stilbene biosynthesis is the conversion of phenylalanine to CA by StlA. Nutrient limitation was recently shown to positively regulate *stlA* gene expression and this was mediated by Lrp, TyrR and σ^S (Lango-Scholey et al. 2013). TyrR was shown to be essential for *stlA* expression (and stilbene production) and, in *Escherichia coli*, TyrR activity is dependent on aromatic amino acids such as phenylalanine (Pittard and Davidson 1991; Lango-Scholey et al. 2013). Therefore, the role of TyrR in *Photorhabdus* may be to couple *stlA* expression to the presence of phenylalanine. In contrast whilst both Lrp and σ^S were required for optimal *stlA* expression in *P. luminescens* TTO1 both of these regulators were dispensable for normal stilbene production (Lango-Scholey et al. 2013). This apparent paradox can be simply explained as *Photorhabdus* produces (and secretes) more CA than is required for stilbene production

(Chalabaev et al. 2008). Therefore, a reduction in CA production due to lower levels of *stlA* expression does not necessarily have to result in a reduction in stilbene production. Interestingly, phenylalanine cannot be used as a carbon source by *Photorhabdus* whilst CA can be assimilated through the activity of the Hca/Mhp pathway (Chalabaev et al. 2008). Therefore, the induction of *stlA* expression following nutrient limitation will, in addition to allowing stilbene production, facilitate optimal nutrient scavenging by *Photorhabdus* in the nutrient limited batch environment of the insect cadaver.

5.2.3 Bioluminescence

Bacterial bioluminescence has been well studied in marine bacteria such as *Vibrio fischeri* and *Vibrio harveyi* where light production has been known for a long time to be regulated by AHL-based quorum-sensing pathways (Meighen 1991). However, *Photorhabdus*, the only known terrestrial bioluminescent bacterium, does not have any AHL-based regulatory circuits. All bioluminescent bacteria have a *lux* operon that contains the 5 genes required for light production, *luxCDABE*. The *luxA* and *luxB* genes encode the light-producing enzyme luciferase and the *luxC* and *luxD* genes encode a fatty acid reductase that produces the aldehyde substrate of luciferase. Light production by *Photorhabdus* increases significantly during the post-exponential phase of bacterial growth suggesting that bioluminescence may have a role during the mutualistic association with the nematode (Schmidt et al. 1989). However, the role and regulation of bioluminescence in *Photorhabdus* is unclear. The expression of the *lux* operon in *P. luminescens* TTO1 is reduced in the *uvrY* mutant. However, the reduction in light production in the *uvrY* mutant can be partially rescued by the exogenous addition of AI-2 to the culture medium and this has been shown to be mediated by the CsrA/*csrB* regulatory system (Krin et al. 2008). Therefore, there may be a significant post-transcriptional component to the regulation of light production in *Photorhabdus*. Interestingly early reports, based on Northern blotting, suggested that the *lux* operon was expressed to the same level in both primary and secondary variants although the level of light production was greatly reduced in the secondary variant (Wang and Dowds 1991). In addition to employing systems based on small regulatory RNA molecules, post-transcriptional regulation can be mediated by the availability of substrate. Acylated urea metabolites (phurealipids) are produced by *Photorhabdus* and have been shown to be antagonists of insect juvenile hormone epoxide hydrolase. Production of phurealipids is dependent on a carbamoyltransferase encoded by *pliA* and disruption of *pliA* completely blocks the production of phurealipids (Nollmann et al. 2015). The substrate for PliA is a fatty acid-derived aldehyde and in independent (and unpublished) experiments mutations in *pliA* were also shown to result in hyper-bioluminescence (Lea Lango-Schooley and David J. Clarke, unpublished data). Therefore, it is possible that bioluminescence may be controlled, at least in part, by limiting the availability of the fatty acid-derived aldehydes required by the light-producing enzyme, luciferase.

6 Conclusions

P. luminescens TTO1 employs a complex regulatory network to control secondary metabolism during a life cycle that involves significant functional heterogeneity through phenotypic and phase variation and contrasting interactions with different invertebrate hosts. The topology of this network is such that global and pathway-specific regulators interact with primary metabolism to temporally regulate secondary metabolism, including bioluminescence and the production of small bioactive metabolites (see Fig. 4). The secondary metabolites produced during the post-exponential phase of *Photorhabus* growth have been linked with a role in the mutualistic association with the nematode. Nutrient limitation (through the production of the alarmone (p)ppGpp) has been shown to be a major environmental factor that controls secondary metabolism in *Photorhabdus*. The production of (p)ppGpp is an internal signal that coordinates the transduction of nutrient limitation into a global regulatory response that, directly or indirectly and through the activities of a number of regulators, initiates the transition from pathogenicity to mutualism. Further unravelling of the topology of these regulatory pathways remains an important ambition for a better understanding of the tripartite interaction between *Photorhabdus*, the insect and the nematode.

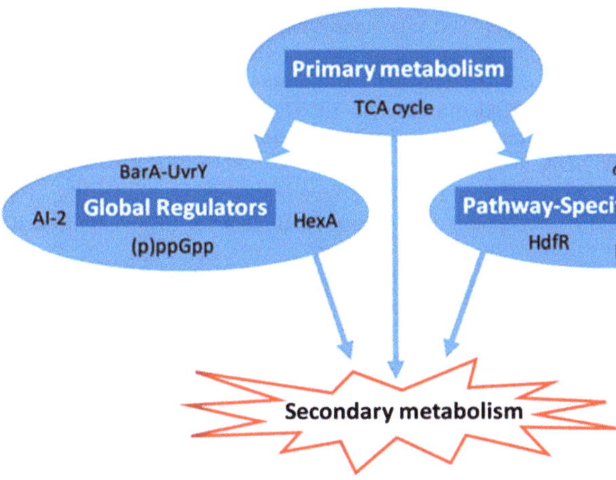

Fig. 4 Regulatory network controlling secondary metabolism in *P. luminescens* TTO1

References

Ackermann M (2015) A functional perspective on phenotypic heterogeneity in microorganisms. Nat Rev Microbiol 13:497–508. doi:10.1038/nrmicro3491

Ahn J-Y, Lee J-Y, Yang E-J et al (2013) Mosquitocidal activity of anthraquinones isolated from symbiotic bacteria *Photorhabdus* of entomopathogenic nematode. J Asia-Pacific Entomol 16:317–320. doi:10.1016/j.aspen.2013.04.005

Akhurst RJ (1980) Morphological and functional dimorphism in *Xenorhabdus* spp., bacteria symbiotically associated with the insect pathogenic nematodes neoaplectana and heterorhabditis. J Gen Microbiol 121:303–309

Babitzke P, Romeo T (2007) CsrB sRNA family: sequestration of RNA-binding regulatory proteins. Curr Opin Microbiol 10:156–163. doi:10.1016/j.mib.2007.03.007

Bager R, Roghanian M, Gerdes K, Clarke DJ (2016) Alarmone (p)ppGpp regulates the transition from pathogenicity to mutualism in *Photorhabdus luminescens*. Mol Microbiol 100:735–747. doi:10.1111/mmi.13345

Bian X, Plaza A, Zhang Y, Müller R (2012) Luminmycins A-C, cryptic natural products from *Photorhabdus luminescens* identified by heterologous expression in *Escherichia coli*. J Nat Prod 75:1652–1655. doi:10.1021/np300444e

Blackburn D, Wood PL, Burk TJ et al (2016) Evolution of virulence in *Photorhabdus* spp., entomopathogenic nematode symbionts. Syst Appl Microbiol. doi:10.1016/j.syapm.2016.02.003

Bode HB (2009) Entomopathogenic bacteria as a source of secondary metabolites. Curr Opin Chem Biol 13:224–230. doi:10.1016/j.cbpa.2009.02.037

Bode HB, Reimer D, Fuchs SW et al (2012) Determination of the absolute configuration of peptide natural products by using stable isotope labeling and mass spectrometry. Chemistry 18:2342–2348. doi:10.1002/chem.201103479

Bode HB, Brachmann AO, Jadhav KB et al (2015) Structure elucidation and activity of Kolossin A, the D-/L-pentadecapeptide product of a giant nonribosomal peptide synthetase. Angew Chem Int Ed Engl 54:10352–10355. doi:10.1002/anie.201502835

Boemare NE, Akhurst RJ (1988) Biochemical and physiological characterization of colony form variants in *Xenorhabdus* spp. (enterobacteriaceae). J Gen Microbiol 134:751–761

Bowden SD, Eyres A, Chung JCS et al (2013) Virulence in *Pectobacterium atrosepticum* is regulated by a coincidence circuit involving quorum sensing and the stress alarmone, (p)ppGpp. Mol Microbiol 90:457–471. doi:10.1111/mmi.12369

Bowen DJ, Ensign JC (2001) Isolation and characterization of intracellular protein inclusions produced by the entomopathogenic bacterium *Photorhabdus luminescens*. Appl Environ Microbiol 67:4834–4841. doi:10.1128/AEM.67.10.4834-4841.2001

Brachmann AO, Joyce SA, Jenke-Kodama H et al (2007) A type II polyketide synthase is responsible for anthraquinone biosynthesis in *Photorhabdus luminescens*. ChemBioChem 8:1721–1728. doi:10.1002/cbic.200700300

Brachmann AO, Kirchner F, Kegler C et al (2012a) Triggering the production of the cryptic blue pigment indigoidine from *Photorhabdus luminescens*. J Biotechnol 157:96–99. doi:10.1016/j.jbiotec.2011.10.002

Brachmann AO, Reimer D, Lorenzen W et al (2012b) Reciprocal cross talk between fatty acid and antibiotic biosynthesis in a nematode symbiont. Angew Chem Int Ed Engl 51:12086–12089. doi:10.1002/anie.201205384

Brachmann AO, Brameyer S, Kresovic D et al (2013) Pyrones as bacterial signaling molecules. Nat Chem Biol 9:573–578. doi:10.1038/nchembio.1295

Brock DA, Douglas TE, Queller DC, Strassmann JE (2011) Primitive agriculture in a social amoeba. Nature 469:393–396. doi:10.1038/nature09668

Brown A, Fernández IS, Gordiyenko Y, Ramakrishnan V (2016) Ribosome-dependent activation of stringent control. Nature. doi:10.1038/nature17675

Camacho MI, Alvarez AF, Chavez RG et al (2015) Effects of the global regulator CsrA on the BarA/UvrY two-component signaling system. J Bacteriol 197:983–991. doi:10.1128/JB.02325-14

Cerenius L, Söderhäll K (2004) The prophenoloxidase-activating system in invertebrates. Immunol Rev 198:116–126

Cerenius L, Lee BL, Söderhäll K (2008) The proPO-system: pros and cons for its role in invertebrate immunity. Trends Immunol 29:263–271. doi:10.1016/j.it.2008.02.009

Chalabaev S, Turlin E, Bay S et al (2008) Cinnamic acid, an autoinducer of its own biosynthesis, is processed via Hca enzymes in *Photorhabdus luminescens*. Appl Environ Microbiol 74:1717–1725. doi:10.1128/AEM.02589-07

Ciche TA, Ensign JC (2003) For the insect pathogen *Photorhabdus luminescens*, which end of a nematode is out? Appl Environ Microbiol 69:1890–1897

Ciche TA, Kim KS, Kaufmann-Daszczuk B et al (2008) Cell Invasion and Matricide during *Photorhabdus luminescens* Transmission by *Heterorhabditis bacteriophora* Nematodes. Appl Environ Microbiol 74:2275–2287. doi:10.1128/AEM.02646-07

Clarke DJ (2008) *Photorhabdus*: a model for the analysis of pathogenicity and mutualism. Cell Microbiol 10:2159–2167. doi:10.1111/j.1462-5822.2008.01209.x

Clarke DJ (2014) The genetic basis of the symbiosis between *Photorhabdus* and its invertebrate hosts. Adv Appl Microbiol 88:1–29. doi:10.1016/B978-0-12-800260-5.00001-2

Clarke DJ, Dowds BC (1995) Virulence mechanisms of *Photorhabdus* sp. strain K122 toward wax moth larvae. J Invertebr Pathol 66:149–155

Crawford JM, Kontnik R, Clardy J (2010) Regulating alternative lifestyles in entomopathogenic bacteria. Curr Biol 20:69–74. doi:10.1016/j.cub.2009.10.059

Crawford JM, Mahlstedt SA, Malcolmson SJ et al (2011) Dihydrophenylalanine: a prephenate-derived *Photorhabdus luminescens* antibiotic and intermediate in dihydrostilbene biosynthesis. Chem Biol 18:1102–1112. doi:10.1016/j.chembiol.2011.07.009

Crawford JM, Portmann C, Zhang X et al (2012) Small molecule perimeter defense in entomopathogenic bacteria. Proc Natl Acad Sci 109:10821–10826. doi:10.1073/pnas.1201160109

Davies JA, Harrison JJ, Marques LLR et al (2007) The GacS sensor kinase controls phenotypic reversion of small colony variants isolated from biofilms of *Pseudomonas aeruginosa* PA14. FEMS Microbiol Ecol 59:32–46. doi:10.1111/j.1574-6941.2006.00196.x

Derzelle S, Duchaud E, Kunst F et al (2002) Identification, characterization, and regulation of a cluster of genes involved in carbapenem biosynthesis in *Photorhabdus luminescens*. Appl Environ Microbiol 68:3780–3789

Derzelle S, Ngo S, Turlin E, et al (2004) AstR-AstS, a new two-component signal transduction system, mediates swarming, adaptation to stationary phase and phenotypic variation in *Photorhabdus luminescens*. Microbiology (Reading, Engl) 150:897–910. doi:10.1099/mic.0.26563-0

Dubnau D, Losick R (2006) Bistability in bacteria. Mol Microbiol 61:564–572. doi:10.1111/j.1365-2958.2006.05249.x

Dudnik A, Bigler L, Dudler R (2013) Heterologous expression of a *Photorhabdus luminescens* syrbactin-like gene cluster results in production of the potent proteasome inhibitor glidobactin A. Microbiol Res 168:73–76. doi:10.1016/j.micres.2012.09.006

Easom CA, Clarke DJ (2012) HdfR is a regulator in *Photorhabdus luminescens* that modulates metabolism and symbiosis with the nematode *Heterorhabditis*. Environ Microbiol 14:953–966. doi:10.1111/j.1462-2920.2011.02669.x

Fenton A, Magoolagan L, Kennedy Z, Spencer KA (2011) Parasite-induced warning coloration: a novel form of host manipulation. Anim Behav 81:417–422. doi:10.1016/j.anbehav.2010.11.010

Ffrench-Constant R, Waterfield N, Daborn P et al (2003) *Photorhabdus*: towards a functional genomic analysis of a symbiont and pathogen. FEMS Microbiol Rev 26:433–456

Fu J, Bian X, Hu S et al (2012) Full-length RecE enhances linear-linear homologous recombination and facilitates direct cloning for bioprospecting. Nat Biotechnol 30:440–446. doi:10.1038/nbt.2183

Gaca AO, Colomer-Winter C, Lemos JA (2015a) Many means to a common end: the intricacies of (p)ppGpp metabolism and its control of bacterial homeostasis. J Bacteriol 7:1146–1156. doi:10.1128/JB.02577-14

Gaca AO, Kudrin P, Colomer-Winter C et al (2015b) From (p)ppGpp to (pp)pGpp: characterization of regulatory effects of pGpp synthesized by the small alarmone synthetase of *Enterococcus faecalis*. J Bacteriol 197:2908–2919. doi:10.1128/JB.00324-15

Goodrich-Blair H, Clarke DJ (2007) Mutualism and pathogenesis in *Xenorhabdus* and *Photorhabdus*: two roads to the same destination. Mol Microbiol 64:260–268. doi:10.1111/j.1365-2958.2007.05671.x

Han R, Ehlers RU (2001) Effect of *Photorhabdus luminescens* phase variants on the in vivo and in vitro development and reproduction of the entomopathogenic nematodes *Heterorhabditis bacteriophora* and *Steinernema carpocapsae*. FEMS Microbiol Ecol 35:239–247

Hauryliuk V, Atkinson GC, Murakami KS et al (2015) Recent functional insights into the role of (p)ppGpp in bacterial physiology. Nature Rev Microbiol 13:298–309. doi:10.1038/nrmicro3448

Heeb S, Haas D (2001) Regulatory roles of the GacS/GacA two-component system in plant-associated and other gram-negative bacteria. Mol Plant Microbe Interact 14:1351–1363. doi:10.1094/MPMI.2001.14.12.1351

Hesketh A, Chen WJ, Ryding J et al (2007) The global role of ppGpp synthesis in morphological differentiation and antibiotic production in *Streptomyces coelicolor* A3(2). Genome Biol 8: R161. doi:10.1186/gb-2007-8-8-r161

Hirano T, Beck DAC, Demuth DR et al (2012) Deep sequencing of *Porphyromonas gingivalis* and comparative transcriptome analysis of a LuxS mutant. Front Cell Infect Microbiol 2:79. doi:10.3389/fcimb.2012.00079

Hu K, Webster JM (2000) Antibiotic production in relation to bacterial growth and nematode development in Photorhabdus-Heterorhabditis infected Galleria mellonella larvae. FEMS Microbiol Lett 189:219–223

Jesudhasan PR, Cepeda ML, Widmer K et al (2010) Transcriptome analysis of genes controlled by *luxS*/autoinducer-2 in *Salmonella enterica* serovar Typhimurium. Foodborne Pathog Dis 7:399–410. doi:10.1089/fpd.2009.0372

Jones RS, Fenton A, Speed MP (2016) "Parasite-induced aposematism" protects entomopathogenic nematode parasites against invertebrate enemies. Behav Ecol 27:645–651. doi:10.1093/beheco/arv202

Joyce SA, Clarke DJ (2003) A *hexA* homologue from *Photorhabdus* regulates pathogenicity, symbiosis and phenotypic variation. Mol Microbiol 47:1445–1457

Joyce SA, Brachmann AO, Glazer I et al (2008) Bacterial biosynthesis of a multipotent stilbene. Angew Chem Int Ed Engl 47:1942–1945. doi:10.1002/anie.200705148

Joyce SA, Lango L, Clarke DJ (2011) The regulation of secondary metabolism and mutualism in the insect pathogenic bacterium *Photorhabdus luminescens*. Adv Appl Microbiol 76:1–25. doi:10.1016/B978-0-12-387048-3.00001-5

Kanost MR, Jiang H, Yu X-Q (2004) Innate immune responses of a lepidopteran insect, *Manduca sexta*. Immunol Rev 198:97–105

Kendall MM, Rasko DA, Sperandio V (2007) Global effects of the cell-to-cell signaling molecules autoinducer-2, autoinducer-3, and epinephrine in a *luxS* mutant of enterohemorrhagic *Escherichia coli*. Infect Immun 75:4875–4884. doi:10.1128/IAI.00550-07

Kontnik R, Crawford JM, Clardy J (2010) Exploiting a global regulator for small molecule discovery in *Photorhabdus luminescens*. ACS Chem Biol 5:659–665. doi:10.1021/cb100117k

Krin E, Chakroun N, Turlin E et al (2006) Pleiotropic role of quorum-sensing autoinducer 2 in *Photorhabdus luminescens*. Appl Environ Microbiol 72:6439–6451. doi:10.1128/AEM.00398-06

Krin E, Derzelle S, Bedard K et al (2008) Regulatory role of UvrY in adaptation of *Photorhabdus luminescens* growth inside the insect. Environ Microbiol 10:1118–1134. doi:10.1111/j.1462-2920.2007.01528.x

Lango L, Clarke DJ (2010) A metabolic switch is involved in lifestyle decisions in *Photorhabdus luminescens*. Mol Microbiol 77:1394–1405. doi:10.1111/j.1365-2958.2010.07300.x

Lango-Scholey L, Brachmann AO, Bode HB, Clarke DJ (2013) The expression of *stlA* in *Photorhabdus luminescens* is controlled by nutrient limitation. PLoS ONE 8:e82152. doi:10.1371/journal.pone.0082152.s004

Lapouge K, Schubert M, Allain FHT, Haas D (2007) Gac/Rsm signal transduction pathway of γ-proteobacteria: from RNA recognition to regulation of social behaviour. Mol Microbiol 67:241–253. doi:10.1111/j.1365-2958.2007.06042.x

Liu MY, Romeo T (1997) The global regulator CsrA of *Escherichia coli* is a specific mRNA-binding protein. J Bacteriol 179:4639–4642

Meighen EA (1991) Molecular biology of bacterial bioluminescence. Microbiol Rev 55:123–142

Meslet-Cladiere LM, Pimenta A, Duchaud E et al (2004) In vivo expression of the mannose-resistant fimbriae of *Photorhabdus temperata* K122 during insect infection. J Bacteriol 186:611–622

Miller MB, Bassler BL (2001) Quorum sensing in bacteria. Annu Rev Microbiol 55:165–199

Nelson LK, Stanton MM, Elphinstone REA et al (2010) Phenotypic diversification in vivo: *Pseudomonas aeruginosa gacS*- strains generate small colony variants in vivo that are distinct from in vitro variants. Microbiology (Reading, Engl) 156:3699–3709. doi:10.1099/mic.0.040824-0

Ng W-L, Bassler BL (2009) Bacterial quorum-sensing network architectures. Annu Rev Genet 43:197–222. doi:10.1146/annurev-genet-102108-134304

Nollmann FI, Heinrich AK, Brachmann AO et al (2015) A *Photorhabdus* natural product inhibits insect juvenile hormone epoxide hydrolase. ChemBioChem 16:766–771. doi:10.1002/cbic.201402650

Park HB, Crawford JM (2015) Lumiquinone A, an α-aminomalonate-derived aminobenzoquinone from *Photorhabdus luminescens*. J Nat Prod 78:1437–1441. doi:10.1021/np500974f

Pereira CS, Thompson JA, Xavier KB (2013) AI-2-mediated signalling in bacteria. FEMS Microbiol Rev 37:156–181. doi:10.1111/j.1574-6976.2012.00345.x

Pernestig A-K, Georgellis D, Romeo T et al (2003) The *Escherichia coli* BarA-UvrY two-component system is needed for efficient switching between glycolytic and gluconeogenic carbon sources. J Bacteriol 185:843–853

Pittard AJ, Davidson BE (1991) TyrR protein of *Escherichia coli* and its role as repressor and activator. Mol Microbiol 5:1585–1592. doi:10.1111/j.1365-2958.1991.tb01904.x

Romeo T (1998) Global regulation by the small RNA-binding protein CsrA and the non-coding RNA molecule CsrB. Mol Microbiol 29:1321–1330

Romero D, Traxler MF, López D, Kolter R (2011) Antibiotics as signal molecules. Chem Rev 111:5492–5505. doi:10.1021/cr2000509

Ross W, Vrentas CE, Sanchez-Vazquez P et al (2013) The magic spot: a ppGpp binding site on *E. coli* RNA polymerase responsible for regulation of transcription initiation. Mol Cell 50:420–429. doi:10.1016/j.molcel.2013.03.021

Schmidt TM, Kopecky K, Nealson KH (1989) Bioluminescence of the insect pathogen *Xenorhabdus luminescens*. Appl Environ Microbiol 55:2607–2612

Septer AN, Bose JL, Lipzen A et al (2015) Bright luminescence of *Vibrio fischeri* aconitase mutants reveals a connection between citrate and the Gac/Csr regulatory system. Mol Microbiol 95:283–296. doi:10.1111/mmi.12864

Smigielski AJ, Akhurst RJ, Boemare NE (1994) Phase variation in *Xenorhabdus nematophilus* and *Photorhabdus luminescens*: differences in respiratory activity and membrane energization. Appl Environ Microbiol 60:120–125

Somvanshi VS, Kaufmann-Daszczuk B, Kim K-S et al (2010) *Photorhabdus* phase variants express a novel fimbrial locus, *mad*, essential for symbiosis. Mol Microbiol 77:1021–1038. doi:10.1111/j.1365-2958.2010.07270.x

Somvanshi VS, Sloup RE, Crawford JM et al (2012) A single promoter inversion switches *Photorhabdus* between pathogenic and mutualistic states. Science 337:88–93. doi:10.1126/science.1216641

Stallforth P, Brock DA, Cantley AM et al (2013) A bacterial symbiont is converted from an inedible producer of beneficial molecules into food by a single mutation in the *gacA* gene. Proc Natl Acad Sci 110:14528–14533. doi:10.1073/pnas.1308199110

Stein ML, Beck P, Kaiser M et al (2012) One-shot NMR analysis of microbial secretions identifies highly potent proteasome inhibitor. Proc Natl Acad Sci 109:18367–18371. doi:10.1073/pnas.1211423109

Strauch O, Ehlers RU (1998) Food signal production of *Photorhabdus luminescens* inducing the recovery of entomopathogenic nematodes *Heterorhabditis* spp. in liquid culture. Appl Microbiol Biotechnol 50:369–374. doi:10.1007/s002530051306

Takeuchi K, Kiefer P, Reimmann C et al (2009) Small RNA-dependent expression of secondary metabolism is controlled by Krebs cycle function in *Pseudomonas fluorescens*. J Biol Chem 284:34976–34985. doi:10.1074/jbc.M109.052571

Takeuchi K, Yamada K, Haas D (2012) ppGpp controlled by the Gac/Rsm regulatory pathway sustains biocontrol activity in *Pseudomonas fluorescens* CHA0. Mol Plant Microbe Interact 25:1440–1449. doi:10.1094/MPMI-02-12-0034-R

Theodore CM, King JB, You J, Cichewicz RH (2012) Production of cytotoxic glidobactins/luminmycins by *Photorhabdus asymbiotica* in liquid media and live crickets. J Nat Prod 75:2007–2011. doi:10.1021/np300623x

Traxler MF, Zacharia VM, Marquardt S et al (2010) Discretely calibrated regulatory loops controlled by ppGpp partition gene induction across the "feast to famine" gradient in *Escherichia coli*. Mol Microbiol 79:830–845. doi:10.1111/j.1365-2958.2010.07498.x

Turlin E, Pascal G, Rousselle J-C et al (2006) Proteome analysis of the phenotypic variation process in *Photorhabdus luminescens*. Proteomics 6:2705–2725. doi:10.1002/pmic.200500646

Veening J-W, Smits WK, Kuipers OP (2008) Bistability, epigenetics, and bet-hedging in bacteria. Annu Rev Microbiol 62:193–210. doi:10.1146/annurev.micro.62.081307.163002

Vercruysse M, Fauvart M, Jans A et al (2011) Stress response regulators identified through genome-wide transcriptome analysis of the (p)ppGpp-dependent response in *Rhizobium etli*. Genome Biol 12:R17. doi:10.1186/gb-2011-12-2-r17

Wang H, Dowds B (1991) Molecular cloning and characterization of the *lux* genes from the secondary form of *Xenorhabdus luminescens*, K122. Biochem Soc Trans 20:68S

Waterfield NR, Ciche T, Clarke D (2009) *Photorhabdus* and a host of hosts. Annu Rev Microbiol 63:557–574. doi:10.1146/annurev.micro.091208.073507

Watson RJ, Joyce SA, Spencer GV, Clarke DJ (2005) The *exbD* gene of *Photorhabdus temperata* is required for full virulence in insects and symbiosis with the nematode *Heterorhabditis*. Mol Microbiol 56:763–773. doi:10.1111/j.1365-2958.2005.04574.x

Weilbacher T, Suzuki K, Dubey AK et al (2003) A novel sRNA component of the carbon storage regulatory system of *Escherichia coli*. Mol Microbiol 48:657–670

Williams JS, Thomas M, Clarke DJ (2005) The gene *stlA* encodes a phenylalanine ammonia-lyase that is involved in the production of a stilbene antibiotic in *Photorhabdus luminescens* TT01. Microbiology (Reading, Engl) 151:2543–2550. doi:10.1099/mic.0.28136-0

You J, Liang S, Cao L et al (2006) Nutritive significance of crystalline inclusion proteins of *Photorhabdus luminescens* in *Steinernema* nematodes. FEMS Microbiol Ecol 55:178–185. doi:10.1111/j.1574-6941.2005.00015.x

Zuo Y, Wang Y, Steitz TA (2013) The mechanism of *E. coli* RNA polymerase regulation by ppGpp is suggested by the structure of their complex. Mol Cell 50:430–436. doi:10.1016/j.molcel.2013.03.020

Quorum Sensing and LuxR Solos in *Photorhabdus*

Sophie Brameyer and Ralf Heermann

Abstract Bacterial communication via small diffusible molecules to mediate group-coordinated behaviour is commonly referred to as 'quorum sensing'. The prototypical quorum sensing system of Gram-negative bacteria consists of a LuxI-type autoinducer synthase that produces acyl-homoserine lactones (AHLs) as signals and a LuxR-type receptor that detects the AHLs to control expression of specific genes. However, many bacteria possess LuxR homologs but lack a cognate LuxI-type AHL-synthase. Those LuxR-type receptors are designated as 'LuxR orphans' or 'solos'. Entomopathogenic bacteria of the genus *Photorhabdus* all harbour a large number of LuxR solos, more than any other bacteria examined so far. Two novel quorum sensing systems were found to regulate cell clumping in *Photorhabdus* and therefore affect pathogenicity. In *Photorhabdus luminescens* and *Photorhabdus temperata* the LuxR solo PluR senses α-pyrones named 'photopyrones' instead of AHLs, which are produced by the pyrone synthase PpyS. In contrast, *Photorhabdus asymbiotica*, a closely related insect and human pathogen, has the PluR homolog PauR, which senses dialkylresorcinols produced by the DarABC pathway to regulate pathogenicity. All three *Photorhabdus* species harbour at least one LuxR solo with an intact AHL-binding motif, which might also allow sensing of exogenous AHLs. However, the majority of the LuxR solos in all *Photorhabdus* species have a PAS4 signal-binding domain. These receptors are assumed to detect eukaryotic compounds and are proposed to be involved in host sensing. Overall, because of the large number of LuxR solos they encode, bacteria of the genus *Photorhabdus* are ideal candidates to study and to identify novel bacterial communication networks.

S. Brameyer · R. Heermann (✉)
Biozentrum, Bereich Mikrobiologie, Ludwig-Maximilians-Universität München,
Großhaderner Str. 2-4, 82152 Martinsried/München, Germany
e-mail: heermann@lmu.de

Current Topics in Microbiology and Immunology (2017) 402:103–120
DOI 10.1007/82_2016_28

Published Online: 14 November 2016

Contents

1 Quorum Sensing in Gram-Negative Bacteria

Bacteria occupy complex environments in nature and often live in close association with other organisms. In order to survive in complex mixed communities, bacteria constantly need to monitor and communicate with their surroundings and to adapt their behaviour accordingly. To process external information and to generate an intracellular response, bacteria have developed several systems, which act either at single cell level, like chemotaxis or two-component systems, or at the population level, like quorum-sensing systems. Regulation of bacterial behaviour in a population density-dependent manner via the use of small signalling molecules is referred to as 'quorum sensing' (QS) (Nealson and Hastings 1979). QS pathways are widespread among microorganisms and several processes are regulated via QS systems in bacteria, such as bioluminescence, antibiotic production, virulence, biofilm formation, motility, and sporulation (Waters and Bassler 2005). The major signalling molecules used for QS can be classified into two different groups among Gram-negative and Gram-positive bacteria. Typically, Gram-negative bacteria use small diffusible molecules, the acyl-homoserine lactones or 'AHLs' derived from fatty acids, whereas Gram-positive bacteria use peptide derivatives for QS-dependent communication. However, one exception is the Gram-positive bacterium *Exiguobacterium* sp. isolated from marine water, which was found to use AHLs for quorum sensing (Biswa and Doble 2013). As *Photorhabdus* species are Gram-negative, here we focus on quorum sensing and LuxR-type receptors in Gram-negative bacteria.

The prototypical quorum sensing system of Gram-negative bacteria consists of a LuxI-like autoinducer synthase that produces AHLs as signals and a LuxR-type receptor that detects the AHLs to control expression of specific genes (Waters and Bassler 2005). AHLs are constantly synthesised by LuxI at a low basal level and sensed by the cognate LuxR-like receptor when they exceed a certain threshold concentration. Upon AHL-binding, LuxR binds to the promoter/operator regions of the target genes/operons adapting expression of several genes in response to the cell count (Fig. 1). Thus, bacteria respond to AHLs and adapt bacterial group-behaviour by regulation of gene expression when bacterial cell density reaches a certain level or 'quorum'. LuxI/LuxR-based quorum sensing systems have been intensively studied. The first system was discovered in *Vibrio fischeri* demonstrating that AHLs are used to regulate light production dependent on cell density (Nasser and Reverchon 2006). Furthermore, LuxR-based cell–cell communication is medically relevant as many pathogenic bacteria use these quorum-sensing systems for effective infection (Rutherford and Bassler 2012). Traditionally, it was thought that each LuxR-type receptor requires a cognate LuxI-like homolog producing the AHL-signalling molecule. However, many proteobacterial genomes encode LuxR homologs lacking a cognate LuxI synthase. These LuxR homologs are designated as LuxR 'orphans' (Patankar and González 2009) or LuxR 'solos' (Subramoni and Venturi 2009). They can exist in AHL-producing bacteria besides an entire LuxI/LuxR system or in bacteria that do not produce any AHLs (Fig. 1).

In general, a LuxR-type receptor is composed of two functional domains, a N-terminal signal-binding domain (SBD) and a C-terminal DNA-binding domain (DBD) (Nasser and Reverchon 2006). The DBD contains the conserved "HTH LUXR" helix-turn-helix motif, which is typical for LuxR-type proteins. Furthermore, LuxR-type regulators are usually transcriptional activators. The N-terminal SBD is important for signal binding, binding specificity, and shaping of

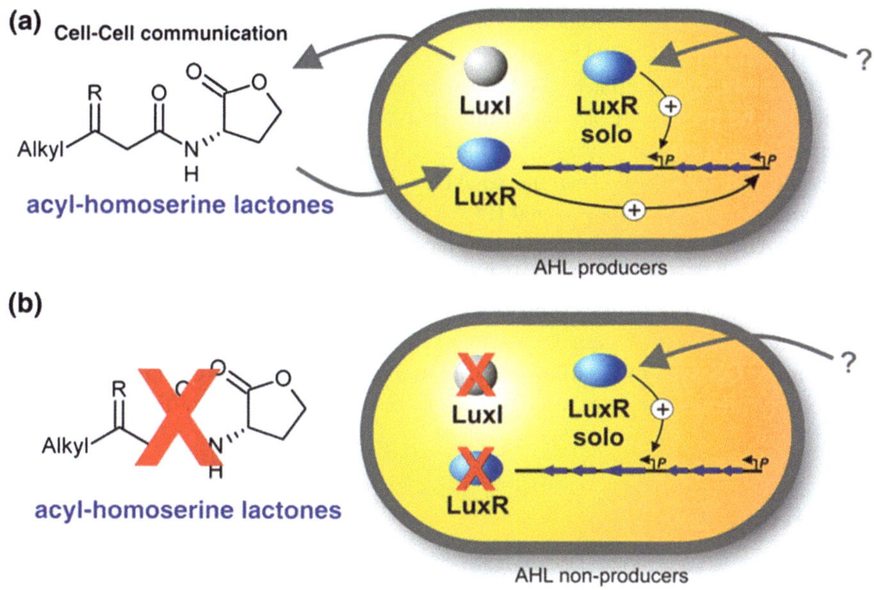

Fig. 1 Quorum sensing and LuxR solos in Gram-negative bacteria. The prototypical quorum sensing system in Gram-negative bacteria is the LuxI/LuxR system, which uses acyl-homoserine lactones (AHLs) for signalling that are produced by LuxI and sensed by the cognate LuxR-type receptor. However, many proteobacteria contain LuxR-type receptors that are not paired with a LuxI-type AHL-synthase. Those so called LuxR orphans or solos occur in AHL-producing (**a**) or AHL non-producing bacteria (**b**). The LuxR-type receptors are drawn in *blue*. R = H/H, H/OH, O

the ligand-binding pocket. Additionally, LuxR-type receptors share a low protein sequence identity (18–25 %). However, nine amino acids are highly conserved within this protein family. The DBD contains three conserved amino acids (E178, L182 and G188, with respect to the LuxR-type receptor TraR of *Agrobacterium tumefaciens*), which are important for DNA-binding. The SBD harbours six conserved amino acids (W57, Y61, D70, P71, W85 and G113, with respect to TraR) that are important for AHL-binding (Fuqua et al. 1996; Patankar and González 2009).

Besides the LuxR-type receptor, the LuxI-type synthase is part of a classical QS system, which synthesises the signalling molecule. LuxI-type synthases are able to synthesise distinct AHLs depending on the precursors, hence the lengths of the acyl moieties of the AHLs can vary between 4 and 18 C-atoms. The third C-atom in the acyl chain can be either a carbonyl group, a hydroxy group or a methylene moiety (Whitehead et al. 2001). These structural differences among the AHLs play a crucial role for signalling specificity of QS LuxR-type receptors in different bacterial species (Kim et al. 2014). Moreover, the signalling molecule is termed autoinducer if transcription of *luxI* is positively regulated via the cognate LuxR-type regulator since this process further boosts AHL synthesis and therefore QS response (Fuqua and Winans 1994). The basic chemical structure of each AHL is identical, but the different AHLs side chains make up certain specificity for the cognate LuxR-type receptor. For that reason, AHLs have been supposed to be one chemical bacterial language, whereas the various derivatives can be treated as different bacterial 'dialects' (Brameyer et al. 2015a).

Furthermore, QS regulatory systems can be more complex as several distinct QS circuits can be present in one bacterium that are inter-connected to each other. For example, *Pseudomonas aeruginosa* harbours three QS systems, the two systems LasI/LasR and RhlI/RhlR respond to AHLs and the third system is dependent on the *Pseudomonas* quinolone signal (PQS) sensed by PqsR and primary synthesised by the *pqsABCD* gene products (Gallagher et al. 2002). Deletion of genes encoding the QS system components reduced *P. aeruginosa* virulence in mice. Moreover, this QS regulatory network controls the expression of different virulence determinants in a hierarchical manner (Martínez 2014; Lee and Zhang 2015). Assuming that a single QS system is sufficient to perceive cell density, a complex QS circuit additionally allows social exploitation and kin recognition (Even-Tov et al. 2016).

2 LuxR Solos

Many proteobacteria posses LuxR 'solos' (Subramoni and Venturi 2009), meaning LuxR-type receptors that are not paired with a cognate LuxI synthase. LuxR solos are supposed to respond to different signals, like exogenous or endogenous AHLs, non-AHLs or eukaryotic signals, and thereby expand the signalling network and influence different cellular processes (Subramoni and Venturi 2009). LuxR-type proteins are mainly restricted to proteobacteria, however recently also few non-proteobacterial sequenced genomes were also found to carry genes that encode LuxR-type proteins. The majority of bacteria contain one or more LuxR solos, either with or without a classical entire QS system (Subramoni et al. 2015). Moreover, only about 26 % of the 265 proteobacterial genomes analysed yet contain genes encoding a complete classical QS circuit (Case et al. 2008). It is therefore obvious that the majority of bacteria use additional and different, yet unexplored, QS systems.

In AHL-producing bacteria, LuxR solos can sense endogenous or exogenous AHLs, and therefore extend the QS regulon to additional target genes. Furthermore, they can respond with different affinities towards specific AHLs and also recognise ligands produced by the surrounding bacterial community. Well-characterised LuxR solos in AHL-producing bacteria are BisR of *Rhizobium leguminosarum* bv. *viciae*, ExpR of *Sinorhizobium meliloti* and QscR of *P. aeruginosa*, which are functional integrated in a resident AHL-mediated QS system (Subramoni and Venturi 2009).

In AHL non-producing bacteria LuxR solos are able to bind exogenous signalling molecules, either AHLs or even other signals produced by eukaryotes, like hormones. These LuxR solos enable bacteria to 'listen' to their bacterial neighbours and benefit from this information to adapt their behaviour according to their environment (Subramoni and Venturi 2009; Hudaiberdiev et al. 2015). In plant-associated bacteria, a new subfamily of LuxR solos was recently described to respond to low molecular weight plant compounds. These LuxR solos are present in beneficial and pathogenic plant-associated bacteria, including members of xanthomonads, rhizobia, agrobacteria and pseudomonads, and are important for plant–bacteria interactions (González and Venturi 2013).

All three *Photorhabdus* species, *Photorhabdus luminescens*, *Photorhabdus temperata* and *Photorhabdus asymbiotica*, contain an unusually high number of genes that encode potential LuxR solos and are therefore assumed to have a huge capacity for cell–cell and/or inter-kingdom communication (Heermann and Fuchs 2008; Wilkinson et al. 2009). *Photorhabdus*-specific LuxR solos share the conserved C-terminal HTH LuxR motif of LuxR-type regulators and are classified based on their distinct N-terminal SBD (Fig. 2). In summary, the presence of the very high number of LuxR solos in all three *Photorhabdus* species probably enable the bacteria to sense diverse signals during their life cycle, and the overall number might reflect the diversity of invertebrate or vertebrate hosts they can infect (Brameyer et al. 2014).

2.1 *AHL-LuxR Solos*

LuxR solos are not necessarily involved in QS processes and may alternatively collect information about the surrounding area (Subramoni and Venturi 2009). The LuxR-type receptor SdiA (Suppressor of cell division inhibitor A) is one prominent example of a LuxR solo that exists in bacteria that do not have a *luxI* gene and

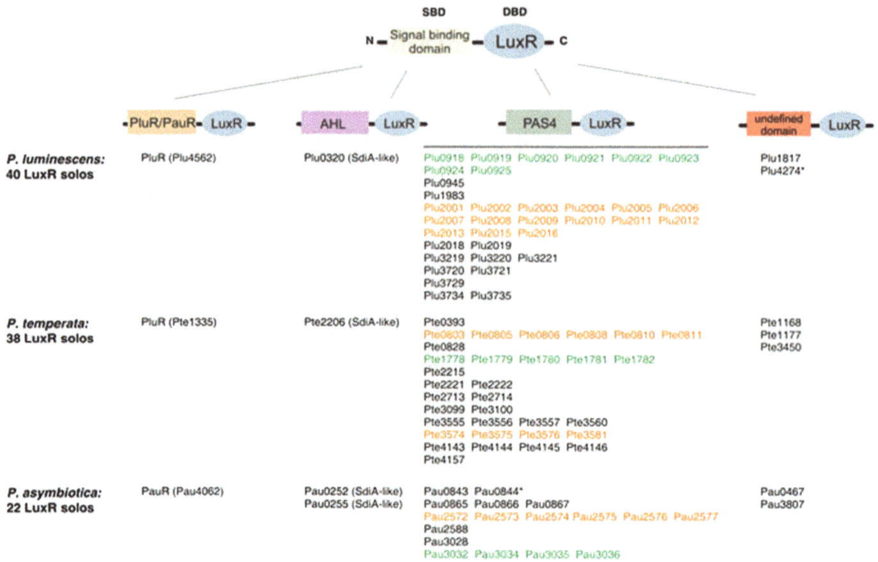

Fig. 2 Domain structure of the several LuxR solos in *P. luminescens, P. asymbiotica* and *P. temperata.* The three types of different LuxR solos and their homologues in *P. luminescens* TT01, *P. temperata* NC19 *and P. asymbiotica* ATCC43949 are shown. The "HTH LUXR" motif (SMART00421) is indicated by a *sphere*, the "Autoind_bind"-domain (PFAM03472) and the "PAS_4"-domain (PFAM08448) by *boxes shaped* in different colours. LuxR solos marked with an asterisk additionally have a predicted N-terminal transmembrane domain (not illustrated). PAS4-LuxR receptors encoded by homologous gene clusters are marked with similar colours (*orange* and *green*) (Brameyer et al. 2014)

therefore do not produce any AHLs, like *Escherichia coli* or *Salmonella enterica* (Ahmer 2004). SdiA homologs are known to detect 'exogenous' AHLs, in other words those produced by neighbouring bacteria enabling these organisms to sense and respond to mixed microbial communities (Michael et al. 2001). In *S. enterica*, SdiA regulates expression of the *rck* and the *srgE* operon in presence of AHLs that were produced by other bacteria (Ahmer et al. 1998; Michael et al. 2001). Since a temperature of 37 °C is essential for recognition of AHLs in *S. enterica*, it is assumed that SdiA is important for sensing the surrounding bacteria when passing the mammalian gut (Ahmer 2004). The SdiA-mediated gene expression in *E. coli* drastically differs from that in *S. enterica*, among others in the temperature optimum. SdiA activates expression of the *ftsQAZ* operon in *E. coli*, which encodes essential cell division proteins (Wang et al. 1991). Furthermore, SdiA is involved in regulating the expression of several genes that are responsible for diverse functions such as metabolism, motility, virulence, survival and defence mechanisms in the presence of AHLs (Kim et al. 2014). Therefore, it is assumed that SdiA-mediated gene expression in *E. coli* is important when changing hosts, especially when the bacteria are exposed to nutrient limitation and other stress outside of a mammalian host (Houdt et al. 2006).

All three *Photorhabdus* species contain at least one AHL-LuxR solo that is homologous to SdiA (Fig. 2). While *P. luminescens* and *P. temperata* each have one SdiA homologue, Plu0320 and Pte2206, respectively, the human pathogen *P. asymbiotica* has two SdiA homologues, Pau0252 and Pau0255 (Brameyer et al. 2014). All SdiA homologues in *Photorhabdus* species have a N-terminal AHL-domain, which is typical for AHL-sensors. Similar to SdiA, the four LuxR solos, Plu0320, Pau0252, Pau0255 and Pte2206 are assumed to detect AHLs as well. Since no *luxI* gene is present in any of the *Photorhabdus* genomes, it is most likely that these LuxR solos respond to exogenously produced AHLs, probably produced by bacteria of the insect gut, which are then sensed by *Photorhabdus* species during the infection process. As the human pathogen *P. asymbiotica* has two putative AHL-sensors, it is most likely that AHL-sensing plays an even more important role for vertebrate than invertebrate infection, putatively by sensing bacteria of the human skin surface. All *Photorhabdus* species have a huge potential to degrade AHLs and therefore to interfere with the QS of other bacteria (Brameyer et al. 2014). They all have several copies of lactonases and acylases, enzymes that hydrolyse the lactone ring and cleave the amide bond of AHLs, respectively. The expression of these different AHL-lactonases and/or AHL-acylases encoding genes could be activated via the respective SdiA homologues in *P. luminescens*, *P. temperata, and P. asymbiotica*, respectively, after sensing external AHLs. Overall, the defence of the dead insect host, which is exposed to competitors in the soil, is crucial for the survival of both *Photorhabdus* bacteria as well as their nematode symbionts and vectors. Thus, 'muzzling' the ambient mixed microbial community of the host and its communication might be an important step for a successful infection process and reproduction. It may also protect the dead insect cadaver from invading saprophytic bacteria as well. It has been proposed that the presence of a second SdiA-homologue in *P. asymbiotica* might be an important step for the successful infection of humans by the otherwise insect pathogen (Brameyer et al. 2014).

2.2 *PluR/PauR-LuxR Solos*

The three LuxR solos PluR (Plu4562) of *P. luminescens*, PauR (Pau4062) of *P. asymbiotica*, and PluR (Pte1335) of *P. temperata* have been previously predicted to be AHL-LuxR solos. However, these LuxR homologs do not sense AHLs, but instead sense photopyrones (PPYs) or dialkylresorcinols (DARs), respectively (Brachmann et al. 2013; Brameyer et al. 2015b). This altered ligand specificity is caused by a modified conserved amino acid motif within the SBD in PluR and PauR (Fig. 3b), which is intact in the SdiA homologues (Figs. 2 and 3a). This altered motif within PluR and PauR is essential for sensing the cognate signalling molecule (Brameyer and Heermann 2015). Amino acid Y61 of AHL-sensing LuxR-type receptors is also conserved in the PluR/PauR motif. This amino acid is known to be involved in binding of the acyl chain of the signalling molecule via hydrophobic interactions, for example in TraR (Churchill and Chen 2011) or LuxR

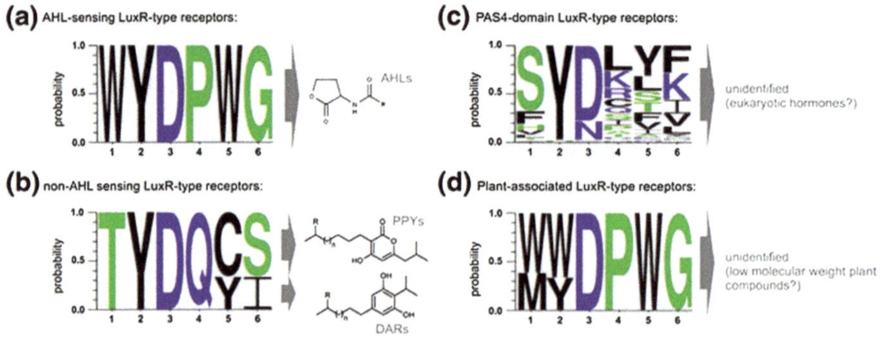

Fig. 3 Conserved amino acid motifs within the signal-binding domain of LuxR-type receptors and their corresponding signalling molecules. **a** Motif of the six conserved amino acid positions in AHL-sensors. Protein sequences of LuxR from *V. fischeri*, TraR from *A. tumefaciens*, SdiA from *Escherichia coli*, QscR and LasR from *Pseudomonas aeruginosa* were used to generate the alignment. **b** Motif of the six conserved amino acids positions of PluR (Plu4562) from *P. luminescens*, PluR (Pte1335) from *P. temperata* and PauR (Pau4062) from *P. asymbiotica* (Brameyer and Heermann 2015). PluR from *P. luminescens* and *P. temperata* sense photopyrones (PPYs) as signalling molecule and PauR senses dialkylresorcinols (DARs) (Brachmann et al. 2013; Brameyer et al. 2015b). **c** Motif of the six conserved amino acid positions of the overall 80 PAS4-LuxR solos in all three *Photorhabdus* species, whereas the corresponding signal molecules are yet unknown, but are possibly eukaryotic hormones (Brameyer and Heermann 2015). **d** Motif of the six conserved amino acids positions of LuxR solos from a subset of plant-associated bacteria possibly sensing low molecular weight plant compounds (Patel et al. 2013)

(Nasser and Reverchon 2006), and is also proposed to be involved in pyrone and dialkylresorcinol binding of PluR and PauR, respectively (Brachmann et al. 2013; Brameyer et al. 2015b). Furthermore, amino acid D70 of AHL-sensors like TraR is known to be important for binding the amide group of *N*-3-oxooctanoyl-L-homoserine lactone (Churchill and Chen 2011). Likewise, D75 (D70 with respect to TraR) of PluR and PauR is deduced to form a hydrogen bond to the hydroxy group attached to the pyrone and the DAR-hydroxy group, respectively (Brachmann et al. 2013; Brameyer et al. 2015b). Therefore, substitution of the conserved amino acid D75 of each PluR and PauR highly decreased recognition of the cognate signalling molecule. Certainly in PluR and PauR, the size and charge of amino acid at position D75 mediates correct signalling molecule binding since substitution against glutamic acid impaired conformation and substitution against asparagine affects binding of PPYD or DAR, respectively. The LuxR solos PluR and PauR are part of novel types of quorum-sensing systems (Brachmann et al. 2013; Brameyer et al. 2015b), which are described below in more detail.

2.3 PAS4-LuxR Solos

The majority of the LuxR solos in *Photorhabdus* contain a N-terminal PAS4 signal-binding domain (Fig. 2). Commonly, PAS (Per-ARNT-Sim) domains are ubiquitous, they are present in archaea, eubacteria and eukarya and are involved in binding of a diverse set of small regulatory molecules either covalent or non-covalent (Hefti et al. 2004). In the fruit fly *Drosophila melangonaster* PAS3 domains have been proven as insect juvenile hormone (JH) receptors (Dubrovsky 2005). The homologous PAS4 domains in *Photorhabdus* are assumed to bind hormone-like molecules and are therefore supposed to be major players in inter-kingdom signalling via the detection of hormone-specific signals from the eukaryotic hosts (Heermann and Fuchs 2008). However, LuxR solos with PAS4 domains of *Photorhabdus* species are assumed to be adapted to signals from the invertebrate hosts (insects and nematodes) or in case of *P. asymbiotica* additionally to vertebrate hosts, especially humans (Brameyer et al. 2014; Wilkinson et al. 2009). The majority of the genes encoding the PAS4-LuxR solos in *Photorhabdus* species are organised in the large gene clusters *plu0918-0925*, *plu2001-2016*, *pau2572-2577*, *pau3032-3036*, *pte0803-0811*, *pte1778-1782* and *pte3574-3581*. The function of this redundancy of PAS4-LuxR solos is still unclear. However, they might respond to a diverse set of eukaryotic signals displaying the wide range of insect hosts they can infect to specifically adapt toxin production and therefore pathogenicity. For that reason, the high redundancy of PAS4-LuxR solos might be a co-evolutional result by adaptation of the bacteria to a broad diversity of insect hosts. This idea is underlined by the fact that several LuxR solos from plant-associated bacteria are known to respond to plant signalling molecules and are therefore assumed to have undergone a co-evolution with the related host plant (Covaceuszach et al. 2013; González and Venturi 2013). *Photorhabdus*-specific LuxR solos with a N-terminal PAS4-domain show diverse variations in the WYDPWG-motif of AHL-sensing LuxR-type regulators, possibly reflecting a higher variety of signals they might sense (Fig. 3c). However, the precise signals that are sensed by PAS4-LuxR solos are yet unknown. The huge diversity of the SBD motifs in PAS4-LuxR solos of *Photorhabdus* and the variations in the conserved amino acid motifs probably gives the bacteria the capacity to respond to a broad range of signals that occur in the different environments. High variability within the WYDPWG-motif in the SBD also occurs in LuxR solos of plant-associated bacteria, which would allow the perception of a broad range of low molecular weight plant compounds rather than AHLs (Fig. 3d). Remarkably, the highly conserved amino acid Y61 of AHL-sensing LuxR-type regulators (with respect to TraR) is present in 97 % of the LuxR solos of *Photorhabdus* species (Brameyer et al. 2014), whereas it is altered in the majority of the LuxR solos from plant-associated bacteria (Fig. 3). This lends support to the idea that different classes of signals can be detected which are specific to either invertebrates or plants.

3 Quorum Sensing in *Photorhabdus*

Photorhabdus bacteria have the capacity to switch readily between different hosts, for example they can colonise either different insects or (in the case of *P. asymbiotica*) humans following their release from their vector *Heterorhabditis* nematodes. Therefore, *Photorhabdus* bacteria constantly need to monitor the local environment and their current host (nematode, insect or man) to appropriately regulate gene expression. Moreover, the behaviour of the whole population needs to be adapted and communicated to be either symbiotic (with its nematode vector) or pathogenic (to either insects or men). Since *Photorhabdus* species do not harbour a LuxI synthase homologue and are therefore unable to produce AHLs, cell–cell communication must be mediated via different signalling molecules. Recently, two LuxR solos sensing non-AHL signalling molecules were described that are used for QS-mediated communication in *Photorhabdus* species both contributing to pathogenicity. One is the LuxR solo PluR from *P. luminescens* and *P. temperata* and the other the LuxR solo PauR from *P. asymbiotica* (Brachmann et al. 2013; Brameyer et al. 2015b). These LuxR solos are not paired with a LuxI synthase.

Notably, *P. luminescens* harbours a LuxS-synthase homolog and has been shown to produce autoinducer-2 (AI-2), which is a furanosyl-borate diesther molecule (Schauder et al. 2001), and acts as a QS molecule as well (Joyce et al. 2011). However, this compound has also been suggested to act as a universal QS signalling molecule in all bacteria (Winzer et al. 2002). LuxS has been demonstrated to influence the carbapenem antibiotic synthesis in *P. luminescens* (Derzelle et al. 2002; Coulthurst et al. 2005). Overall, more than 300 genes have been identified to be under control of LuxS. These genes are involved in metabolism, regulation and general stress response, as well as in pathogenicity. A deletion of *luxS* in *P. luminescens* resulted in reduced biofilm formation and reduced pathogenicity against insects (Krin et al. 2006) However, the detailed molecular mechanisms of LuxS-mediated signalling in *P. luminescens* are still unclear.

3.1 The PpyS/PluR Quorum-Sensing System of P. luminescens *and* P. temperata

The LuxR solo PluR was formerly annotated to contain a putative AHL-binding domain in the N-terminal signal-binding domain (SBD). However, it has been found that this regulator does not recognise AHLs signalling molecules but α-pyrones named 'photopyrones' (PPYs) (Brachmann et al. 2013). The ketosynthase enzyme PpyS is necessary and sufficient for the synthesis of eight different PPYs, depending on the precursors from the fatty acid metabolism. A two-chain biosynthesis step has been described for the synthesis of photopyrones where thioester-activated 9-methyldecanoic acid is first covalently bound to an active site cysteine, and then deprotonated (Kresovic et al. 2015). This results in the formation of a nucleophile, which subsequently attacks the carbonyl carbon of a 5-methyl-3-oxohexanoyl thioester that is formed by the BkdABC pathway (Brachmann et al. 2012) to form a new carbon–carbon (C–C) bond. After an additional deprotonation of the bound intermediate the α-pyrone ring is formed and the respective PPY is released from PpyS (Kresovic et al. 2015). Due to the variability of the first substrate regarding chain length and starting unit, different PPYs are produced. PPYs are produced in standard complex laboratory media, as well as in insect larvae. Furthermore, production of PPYs is dependent on the cell density of a *Photorhabdus* population. Eight α-pyrones (PPYA-PPYH) were isolated from *P. temeperata* spp. *thracensis*, whereas only three (PPYA, PPYB and PPYD) are produced by *P. luminescens*. These three PPYs are all sensed with altered specificity. PluR most specifically senses PPYD with the highest sensitivity in a concentration as low as 3.5 nM, but is also able to sense the other PPYs but only at higher concentrations. It has been proposed that, similar to AHLs, PPY-sensing can be seen as a novel kind of bacterial 'language', whereas the different PPYs can be compared to different bacterial 'dialects' (Brameyer et al. 2015a).

Upon PPY-binding, PluR activates expression of the adjacent *pcfABCDEF* operon. The *pcf* operon encodes a synthesis pathway for a 'clumping' factor named PCF (*Photorhabdus* Clumping Factor) as induction of the operon caused intense cell clumping (Fig. 4a). Insecticidal bioassays using normally harmless *E. coli* cells over-expressing *pcfABCDEF* resulted in high mortality of Greater waxmoth, *Galleria mellonella*, larvae. This led to the conclusion that PCF is a virulence factor that contributes to the high pathogenicity of the bacteria (Brachmann et al. 2013). The entire QS system comprising PluR, the PluR-target operon *pcf*, and PpyS was functionally reconstituted in *E. coli* confirming the nature of this novel cell-cell communication circuit. However, deletion of *pluR* in *P. luminescens* did not result in decreased pathogenicity of the bacteria against insects (Brachmann et al. 2013), but in a decreased symbiotic association with entomopathogenic *Heterorhabditis* nematodes, revealing that cell clumping and therefore QS must also be important for symbiosis with the nematodes (our own unpublished data).

The class of α-pyrones is not unique to *Photorhabdus*, but were correlated to cell–cell communication for the first time with these bacteria. α-pyrones are well-known secondary metabolites in species like pseudomonads (Chu et al. 2002; Kong et al. 2005), fungi (Elbandy et al. 2009) and streptomycetes (Chemler et al. 2012). In streptomycetes pyrones are involved in spore germination (Aoki et al. 2011). Thus, these organisms might use pyrones as signalling molecules beside their other known functions as well.

3.2 The DarABC/PauR Quorum Sensing System *of* P. asymbiotica

The insect and human pathogen *P. asymbiotica* harbours a *pcf* regulon including the LuxR solo PauR that is highly homologous to PluR of *P. luminescens*. However,

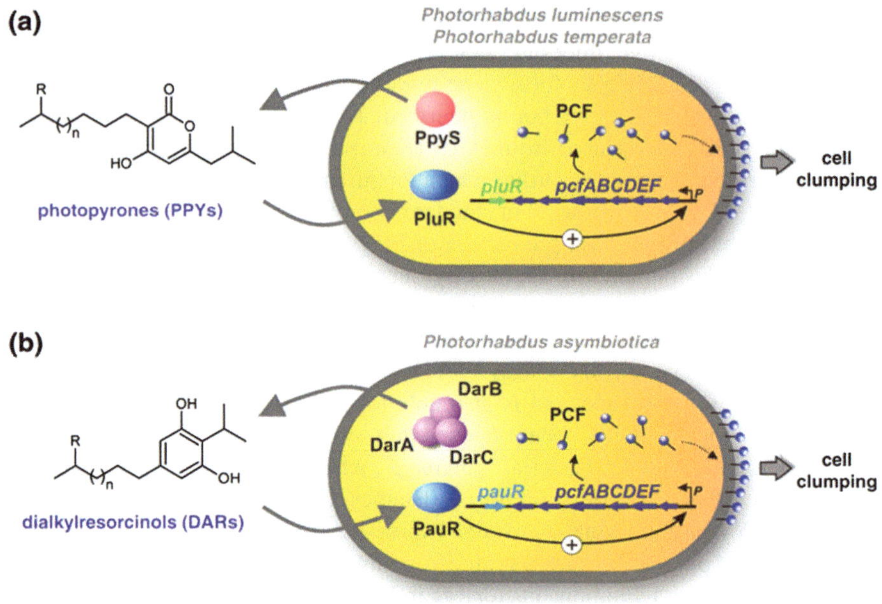

Fig. 4 Quorum sensing systems in *Photorhabdus* species. **a** In *P. luminescens* and *P. temperata* QS to control cell clumping is mediated via photopyrones (PPYs) as signalling molecules, which are produced by the photopyrone synthase PpyS and sensed by the LuxR solo PluR (Brachmann et al. 2013). **b** *P. asymbiotica* uses dialkylresorcinols (DARs) as signalling molecules to control cell clumping, which are synthesised by the DarA/DarB/DarC pathway and sensed by the LuxR solo PauR (Brameyer et al. 2015). The LuxR solos PluR and PauR are both drawn in *blue*. R = H or Me, *n* = 1–4

P. asymbiotica contains neither a PpyS nor a LuxI homologue and does therefore not produce PPYs or AHLs. The signalling molecules sensed by PauR have been identified to belong to the class of dialkylresorcinols (DARs). Upon binding of 2,5-dialkylresorcinol (DAR) to PauR, cell clumping is induced due to expression of the *pcf* operon (Fig. 3b) (Brameyer et al. 2015b). DARs and its biochemical precursors cyclohexanediones (CHDs) are synthesised by enzymes encoded by the *darABC* operon. However, *P. asymbiotica* produces a specific subset of DARs and CHDs depending on the precursors derived from two fatty acids (Fuchs et al. 2013). PauR most specifically recognises DAR at concentrations as low as 5 nM (Brameyer et al. 2015b). Plenty of bacteria possess DarABC homologs, several of them being pathogenic to animals, plants or humans, like various *Neisseria* strains for example (Fuchs et al. 2013). Therefore, the presence of the *darABC* operon and the respective metabolites might be linked to virulence. DAR derivatives have also been described as antibiotics (Joyce et al. 2008), cytotoxins (Kronenwerth et al. 2014), free radical scavengers (Kato et al. 1993), and growth-stimulating factors (Imai et al. 1993). Furthermore, DARs produced by *Pseudomonas* sp. can have antimicrobial properties (Pohanka et al. 2006). Therefore, it might be possible that

DAR-dependent regulation of virulence in general might be more important for colonisation of vertebrates than for invertebrates. It has been suggested that DARs might also be used as QS molecules in many more bacteria, even when they co-occur with LuxR solos. Similar to the PCF of *P. luminescens*, heterologous expression of the *P. asymbiotica pcf* operon in *E. coli* also lead to insect pathogenicity and clumping of normally non-pathogenic *E. coli* cells. In contrast to *pluR* in *P. luminescens*, deletion of *pauR* in *P. asymbiotica* resulted in highly decreased pathogenicity towards *G. mellonella* insect larvae (Brameyer et al. 2015b). For that reason, DAR-mediated QS systems might be a useful and specific target for novel antimicrobials in human pathogenic bacteria.

4 Outlook and Future Perspectives

In the past 10 years several signalling molecules beside AHLs have been implicated in QS-regulated processes, like aryl-HSLs from *Bradyrhizobium* and *Rhodopseudomonas* (Schaefer et al. 2008), photopyrones from *P. luminescens* and *P. temperata* (Brachmann et al. 2013) and dialkylresorcinols from *P. asymbiotica* (Brameyer et al. 2015b). Besides different AHLs, *P. aeruginosa* also uses 2-heptyl-3-hydroxy-4-quinolone (PQS) and the newly identified 2-(2-hydroxyphenyl)-thiazole-4-carbaldehyde (IQS) as languages (Lee et al. 2013). Hence, signalling molecules that are not AHLs appear to play an important role in bacterial communication circuits that have been identified so far, and probably many more await discovery. Moreover, LuxR solos emerge to be more and more important players in cell–cell communication or inter-kingdom signalling as they offer possibilities to use alternative communication molecules to AHLs (Venturi and Ahmer 2015). All three *Photorhabdus* species have an extraordinary high number of LuxR solos, which makes them to optimal model organisms for studying the function of LuxR solos in bacteria. *Photorhabdus*-specific LuxR solos contain different SBDs, which include diverse amino acid motifs at conserved positions compared to AHL-sensing LuxR-type regulators. The diversity of these motifs gives rise to the speculation that signal-binding of all these LuxR solos goes far beyond AHL-signalling as it has been demonstrated for PluR and PauR (Brachmann et al. 2013; Brameyer and Heermann 2015; Brameyer et al. 2015b). Thereby, regulation via LuxR solos is probably important at different steps in the *Photorhabdus* life and infection cycle (Fig. 5). One can only guess the variety of signals perceived by all these LuxR solos and their function in cell–cell communication and inter-kingdom signalling. It will be the goal of future research to unravel the various signalling molecules and correlate them to the specific LuxR solos or amino acid motifs in the signal-binding domain of these proteins. The presence of all those different types of LuxR solos gains first insight into the complexity of the communication network between bacteria among each other and with their hosts. Since most of the LuxR solos that have been investigated so far are involved in regulation of pathogenicity, the homologous receptors or the related

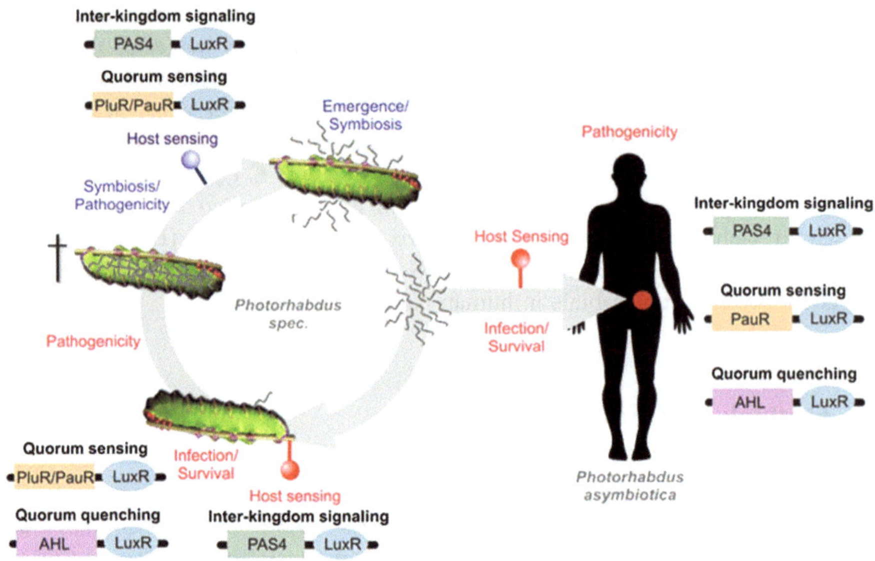

Fig. 5 Overview of the possible role of LuxR solos during life and infection cycle of *Photorhabdus* species. The bacteria colonise the upper gut of heterorhabditis nematodes that invade insect larvae. After release into the insect's hemolymph the bacteria produce several toxins that rapidly kill the prey. After death of the insect host, the bacteria degrade the cadaver and additionally support nematode development. When the cadaver is depleted from nutrients, bacteria and nematode re-associate and leave the carcass in search for a new victim. *P. asymbiotica* can additionally infect humans by inducing systemic and soft tissue infections (*right panel*). The *red* (pathogenic) or *blue* (symbiotic) tags indicate the points within the life cycle where the bacteria switch hosts. Putative involvement of the different LuxR solos (*same colour* code as in Fig. 2) at steps of the infection process or host sensing is indicated at the respective position of the life cycle (see text for details). The figure is modified after Brameyer et al. (2014)

signalling molecules in human pathogens are promising specific drug targets of novel antimicrobials.

Acknowledgments This work was supported by the Deutsche Forschungsgemeinschaft under grant number HE 5247/4-2. We thank Dr. Helge B. Bode and his group for the excellent research cooperation and Dr. Frank Landgraf for critical reading of the manuscript.

References

Ahmer BMM (2004) Cell-to-cell signalling in *Escherichia coli* and *Salmonella enterica*. Mol Microbiol 52:933–945. doi:10.1111/j.1365-2958.2004.04054.x

Ahmer BMM, van Reewuk J, Timmers CD et al (1998) *Salmonella typhimurium* encodes an SdiA homolog, a putative quorum sensor of the LuxR family, that regulates genes on the virulence plasmid. J Bacteriol 1–9

Aoki Y, Matsumoto D, Kawaide H, Natsume M (2011) Physiological role of germicidins in spore germination and hyphal elongation in *Streptomyces coelicolor* A3(2). J Antibiot 64:607–611. doi:10.1038/ja.2011.59

Biswa P, Doble M (2013) Production of acylated homoserine lactone by gram-positive bacteria isolated from marine water. FEMS Microbiol Lett 343:34–41. doi:10.1111/1574-6968.12123

Brachmann AO, Reimer D, Lorenzen W et al (2012) Reciprocal cross talk between fatty acid and antibiotic biosynthesis in a nematode symbiont. Angew Chem Int Ed Engl 51:12086–12089. doi:10.1002/anie.201205384

Brachmann AO, Brameyer S, Kresovic D et al (2013) Pyrones as bacterial signaling molecules. Nat Chem Biol 9:573–578. doi:10.1038/nchembio.1295

Brameyer S, Heermann R (2015) Specificity of signal-binding via non-AHL LuxR-type receptors. PLoS ONE 10:e0124093. doi:10.1371/journal.pone.0124093

Brameyer S, Kresovic D, Bode HB, Heermann R (2014) LuxR solos in *Photorhabdus* species. Front Cell Infect Microbiol 4:166. doi:10.3389/fcimb.2014.00166

Brameyer S, Bode HB, Heermann R (2015a) Languages and dialects: bacterial communication beyond homoserine lactones. Trends Microbiol 23:521–523. doi:10.1016/j.tim.2015.07.002

Brameyer S, Kresovic D, Bode HB, Heermann R (2015b) Dialkylresorcinols as bacterial signaling molecules. Proc Natl Acad Sci U S A 112:572–577. doi:10.1073/pnas.1417685112

Case RJ, Labbate M, Kjelleberg S (2008) AHL-driven quorum-sensing circuits: their frequency and function among the Proteobacteria. ISME J 2:345–349. doi:10.1038/ismej.2008.13

Chemler JA, Buchholz TJ, Geders TW et al (2012) Biochemical and structural characterization of germicidin synthase: analysis of a type III polyketide synthase that employs acyl-ACP as a starter unit donor. J Am Chem Soc 134:7359–7366. doi:10.1021/ja2112228

Chu M, Mierzwa R, Xu L et al (2002) Structure of Sch 419560, a novel alpha-pyrone antibiotic produced by *Pseudomonas fluorescens*. J Antibiot 55:215–218

Churchill MEA, Chen L (2011) Structural basis of acyl-homoserine lactone-dependent signaling. Chem Rev 111:68–85. doi:10.1021/cr1000817

Coulthurst SJ, Barnard AML, Salmond GPC (2005) Regulation and biosynthesis of carbapenem antibiotics in bacteria. Nat Rev Microbiol 3:295–306. doi:10.1038/nrmicro1128

Covaceuszach S, Degrassi G, Venturi V, Lamba D (2013) Structural insights into a novel interkingdom signaling circuit by cartography of the ligand-binding sites of the homologous quorum sensing LuxR-family. IJMS 14:20578–20596. doi:10.3390/ijms141020578

Derzelle S, Duchaud E, Kunst F et al (2002) Identification, characterization, and regulation of a cluster of genes involved in carbapenem biosynthesis in *Photorhabdus luminescens*. Appl Environ Microbiol 68:3780–3789. doi:10.1128/AEM.68.8.3780-3789.2002

Dubrovsky EB (2005) Hormonal cross talk in insect development. Trends Endocrinol Metab 16:6–11. doi:10.1016/j.tem.2004.11.003

Elbandy M, Shinde PB, Hong J et al (2009) Alpha-pyrones and yellow pigments from the sponge-derived fungus *Paecilomyces lilacinus*. Bull Kor Chem Soc 30:188–192

Even-Tov E, Bendori SO, Valastyan J, et al (2016) Social Evolution Selects for Redundancy in Bacterial Quorum Sensing. PLoS Biol 14:e1002386. doi:10.1371/journal.pbio.1002386

Fuchs SW, Bozhüyük KAJ, Kresovic D et al (2013) Formation of 1,3-cyclohexanediones and resorcinols catalyzed by a widely occuring ketosynthase. Angew Chem Int Ed 52:4108–4112. doi:10.1002/anie.201210116

Fuqua WC, Winans SC (1994) A LuxR-LuxI type regulatory system activates *Agrobacterium* Ti plasmid conjugal transfer in the presence of a plant tumor metabolite. J Bacteriol 176:2796–2806

Fuqua C, Winans SC, Greenberg EP (1996) Census and consensus in bacterial ecosystems: the LuxR-LuxI family of quorum-sensing transcriptional regulators. Annu Rev Microbiol 50:727–751. doi:10.1146/annurev.micro.50.1.727

Gallagher LA, McKnight SL, Kuznetsova MS et al (2002) Functions required for extracellular quinolone signaling by *Pseudomonas aeruginosa*. J Bacteriol 184:6472–6480. doi:10.1128/JB.184.23.6472-6480.2002

González JE, Venturi V (2013) A novel widespread interkingdom signaling circuit. Trends Plant Sci 18:167–174. doi:10.1016/j.tplants.2012.09.007

Heermann R, Fuchs TM (2008) Comparative analysis of the *Photorhabdus luminescens* and the *Yersinia enterocolitica* genomes: uncovering candidate genes involved in insect pathogenicity. BMC Genomics 9:40. doi:10.1186/1471-2164-9-40

Hefti MH, Françoijs K-J, de Vries SC, et al (2004) The PAS fold. A redefinition of the PAS domain based upon structural prediction. Eur J Biochem 271:1198–1208

Houdt R, Aertsen A, Moons P et al (2006) N-acyl-L-homoserine lactone signal interception by *Escherichia coli*. FEMS Microbiol Lett 256:83–89. doi:10.1111/j.1574-6968.2006.00103.x

Hudaiberdiev S, Choudhary KS, Vera Alvarez R et al (2015) Census of solo LuxR genes in prokaryotic genomes. Front Cell Infect Microbiol 5:20. doi:10.3389/fcimb.2015.00020

Imai S, Fujioka K, Furihata K et al (1993) Studies on cell growth stimulating substances of low molecular weight. Part 3. Resorcinin, a mammalian cell growth stimulating substance produced by *Cytophaga johnsonae*. J Antibiot 46:1319–1322

Joyce SA, Brachmann AO, Glazer I et al (2008) Bacterial biosynthesis of a multipotent stilbene. Angew Chem Int Ed Engl 47:1942–1945. doi:10.1002/anie.200705148

Joyce SA, Lango L, Clarke DJ (2011) The regulation of secondary metabolism and mutualism in the insect pathogenic bacterium *Photorhabdus luminescens*. In: Laskin Allen, Gadd Geoffrey, Sariaslani Sima (eds) Advances in applied microbiology, vol 76. Academic Press, Burlington, pp 1–25

Kato S, Shindo K, Kawai H et al (1993) Studies on free radical scavenging substances from microorganisms. III. Isolation and structural elucidation of a novel free radical scavenger, resorstatin. J Antibiot 46:1024–1026

Kim T, Duong T, Wu CA et al (2014) Structural insights into the molecular mechanism of *Escherichia coli* SdiA, a quorum-sensing receptor. Acta Crystallogr D Biol Crystallogr 70:694–707. doi:10.1107/S1399004713032355

Kong F, Singh MP, Carter GT (2005) Pseudopyronines A and B, alpha-pyrones produced by a marine *Pseudomonas* sp. F92S91, and evidence for the conversion of 4-hydroxy-alpha-pyrone to 3-furanone. J Nat Prod 68:920–923. doi:10.1021/np050038v

Kresovic D, Schempp F, Cheikh-Ali Z, Bode HB (2015) A novel and widespread class of ketosynthase is responsible for the head-to-head condensation of two acyl moieties in bacterial pyrone biosynthesis. Beilstein J Org Chem 11:1412–1417. doi:10.3762/bjoc.11.152

Krin E, Chakroun N, Turlin E et al (2006) Pleiotropic role of quorum-sensing autoinducer 2 in *Photorhabdus luminescens*. Appl Environ Microbiol 72:6439–6451. doi:10.1128/AEM.00398-06

Kronenwerth M, Brachmann AO, Kaiser M, Bode HB (2014) Bioactive derivatives of isopropylstilbene from mutasynthesis and chemical synthesis. ChemBioChem. doi:10.1002/cbic.201402447

Lee J, Zhang L (2015) The hierarchy quorum sensing network in *Pseudomonas aeruginosa*. Protein Cell 6:26–41. doi:10.1007/s13238-014-0100-x

Lee J, Wu J, Deng Y et al (2013) A cell–cell communication signal integrates quorum sensing and stress response. Nat Chem Biol 9:339–343. doi:10.1038/nchembio.1225

Martínez JL (2014) Interkingdom signaling and its consequences for human health. Virulence 5:243–244. doi:10.4161/viru.28073

Michael BB, Smith JNJ, Swift SS et al (2001) SdiA of *Salmonella enterica* is a LuxR homolog that detects mixed microbial communities. J Bacteriol 183:5733–5742. doi:10.1128/JB.183.19.5733-5742.2001

Nasser W, Reverchon S (2006) New insights into the regulatory mechanisms of the LuxR family of quorum sensing regulators. Anal Bioanal Chem 387:381–390. doi:10.1007/s00216-006-0702-0

Nealson KH, Hastings JW (1979) Bacterial bioluminescence: its control and ecological significance. Microbiol Rev 43:496–518

Patankar AV, González JF (2009) Orphan LuxR regulators of quorum sensing. FEMS Microbiol Rev 33:739–756. doi:10.1111/j.1574-6976.2009.00163.x

Patel HK, Ferrante P, Covaceuszach S et al (2013) The kiwifruit emerging pathogen *Pseudomonas syringae* pv. actinidiae does not produce AHLs but possesses three LuxR solos. PLoS ONE 9: e87862–e87862. doi:10.1371/journal.pone.0087862.s004

Pohanka A, Levenfors J, Broberg A (2006) Antimicrobial dialkylresorcinols from *Pseudomonas* sp. Ki19. J Nat Prod 69:654–657. doi:10.1021/np0600595

Rutherford ST, Bassler BL (2012) Bacterial quorum sensing: its role in virulence and possibilities for its control. Cold Spring Harb Perspect Med 2:a012427–a012427. doi:10.1101/cshperspect.a012427

Schaefer AL, Greenberg EP, Oliver CM et al (2008) A new class of homoserine lactone quorum-sensing signals. Nature 454:595–599. doi:10.1038/nature07088

Schauder S, Shokat K, Surette MG, Bassler BL (2001) The LuxS family of bacterial autoinducers: biosynthesis of a novel quorum-sensing signal molecule. Mol Microbiol 41:463–476

Subramoni S, Florez Salcedo DV, Suarez-Moreno ZR (2015) A bioinformatic survey of distribution, conservation, and probable functions of LuxR solo regulators in bacteria. Front Cell Infect Microbiol 5:16. doi:10.3389/fcimb.2015.00016

Subramoni S, Venturi V (2009) LuxR-family "solos": bachelor sensors/regulators of signalling molecules. Microbiology 155:1377–1385. doi:10.1099/mic.0.026849-0

Venturi V, Ahmer BMM (2015) Editorial: LuxR solos are becoming major players in cell–cell communication in bacteria. Front Cell Infect Microbiol 5:89. doi:10.3389/fcimb.2015.00089

Wang XD, de Boer PA, Rothfield LI (1991) A factor that positively regulates cell division by activating transcription of the major cluster of essential cell division genes of *Escherichia coli*. EMBO J 10:3363–3372

Waters CM, Bassler BL (2005) Quorum sensing: cell-to-cell communication in bacteria. Annu Rev Cell Dev Biol 21:319–346. doi:10.1146/annurev.cellbio.21.012704.131001

Whitehead NA, Barnard AM, Slater H, et al (2001) Quorum-sensing in Gram-negative bacteria. FEMS Microbiol Rev 25:365–404

Wilkinson P, Waterfield NR, Crossman L et al (2009) Comparative genomics of the emerging human pathogen *Photorhabdus asymbiotica* with the insect pathogen *Photorhabdus luminescens*. BMC Genomics 10:302. doi:10.1186/1471-2164-10-302

Winzer K, Hardie KR, Williams P (2002) Bacterial cell-to-cell communication: sorry, can't talk now - gone to lunch! Curr Opin Microbiol 5:216–222

Part III
Insect Immunity

Insect Immunity to Entomopathogenic Nematodes and Their Mutualistic Bacteria

I. Eleftherianos, U. Shokal, S. Yadav, E. Kenney and T. Maldonado

Abstract Entomopathogenic nematodes are important organisms for the biological control of insect pests and excellent models for dissecting the molecular basis of the insect immune response against both the nematode parasites and their mutualistic bacteria. Previous research involving the use of various insects has found distinct differences in the number and nature of immune mechanisms that are activated in response to entomopathogenic nematode parasites containing or lacking their associated bacteria. Recent studies using model insects have started to reveal the identity of certain molecules with potential anti-nematode or antibacterial activity as well as the molecular components that nematodes and their bacteria employ to evade or defeat the insect immune system. Identification and characterization of the genes that regulate the insect immune response to nematode–bacteria complexes will contribute significantly to the development of improved practices to control insects of agricultural and medical importance, and potentially nematode parasites that infect mammals, perhaps even humans.

Contents

I. Eleftherianos (✉) · U. Shokal · S. Yadav · E. Kenney · T. Maldonado
Department of Biological Sciences, The George Washington University,
800 22nd Street NW, Washington DC 20052, USA
e-mail: ioannise@gwu.edu

Current Topics in Microbiology and Immunology (2017) 402:123–156
DOI 10.1007/82_2016_52

Published Online: 08 December 2016

1 Introduction

Entomopathogenic nematodes are ubiquitous natural obligate parasites that infect a wide range of insects. They use the infected insect to complete their life cycle and replicate within the cadaver before they disperse to search for new hosts. The main feature of these parasites is the presence of mutualistic bacteria in the gut of the infective juvenile (IJ) nematodes, these IJ's are unable to feed but can exist outside insect cadaver and are designed to seek out and physically penetrate new insect hosts (Forst et al. 1997). Following entry into the insect host by the IJ, the bacteria are released in the insect open circulatory system (termed hemolymph) where they replicate at high rates while they secrete a battery of toxins, virulence factors and degradative enzymes that target certain host tissues and result in rapid insect death (Waterfield et al. 2009). The bacteria also provide nutrients to the nematodes, they aid nematodes in development and reproduction, and produce antibiotics that inhibit competing microbes in the insect (Eleftherianos 2009). The life cycle of entomopathogenic nematodes can be divided into five distinct stages: (1) penetration into the hemocoel of the potential insect host, (2) recovery into the hemolymph, (3) development to mature stage and reproduction, (4) production of new infective juveniles, and (5) host searching and host acceptance by the infective juveniles (Forst and Clarke 2002). The basis of the success of the parasitic nematodes and their mutualistic bacteria in insect infection is the interaction of each partner of the nematode-bacteria complex with the insect immune system (Eleftherianos et al. 2010b; Castillo et al. 2011).

Entomopathogenic nematodes and their mutualistic bacteria are widely used in biological control regimes and as an alternative to chemical insecticides for the efficient management of agricultural insect pests, mainly lepidopteran larvae that cause severe damage to large quantities of crops (Castagnola and Stock 2014; Shapiro-llan et al. 2012). They also form excellent models for dissecting the molecular and mechanistic basis of nematode parasitism, bacterial virulence, microbial mutualism and elucidating the interaction among these important biological functions (Ffrench-Constant et al. 2007a, b; Goodrich-Blair and Clarke 2007). In addition, the nematode parasites and their associated bacteria are outstanding pathogenic organisms for probing the insect host anti-nematode and antibacterial immune response (Kenney and Eleftherianos 2016). Insects have developed sensitive mechanisms for detecting the presence of microbial infections and activating signaling pathways that control the production of molecules with antimicrobial activity (Hillyer 2016).

Previous work has started to expose the nature and genetic basis of the mechanisms that insects activate in response to entomopathogenic nematodes and their mutualistic bacteria. Here we review the insect immune responses that are directed against insect nematodes parasites and their related bacteria, as well as the host immune genes that are targeted by the pathogens during the infection process. This information is important for understanding the dynamics of interaction of nematode–bacteria complexes with the insect immune system and the factors that determine the outcome of infection. Such knowledge is essential to fully exploit the biological control potential of entomopathogenic nematodes against deleterious insect pests.

2 Entomopathogenic Nematode–Bacteria Complexes

Entomopathogenic nematodes are soil-dwelling, obligate and lethal parasites of insects that are able to search, penetrate, release their mutualistic bacteria, kill and reproduce within the insect cadaver and then emerge as infective juveniles (Dillman et al. 2012). The most well studied entomopathogenic nematodes belong to the order Rhabditida in two families: Heterorhabditidae and Steinernematidae. Entomopathogenic nematodes in the genera *Heterorhabditis* and *Steinernema* have been recovered from all continents except Antarctica (Griffin et al. 1990), and they can infect a broad range of insect species, including important lepidopteran, dipteran, homopteran, and coleopteran pests of cultivated plants (Peters 2013). The main feature of *Heterorhabditis* and *Steinernema* nematodes is their unique mutualistic association with the enteric insect-pathogenic bacteria *Photorhabdus* and *Xenorhabdus*, respectively (Boemare 2002).

Entomopathogenic nematodes have a life cycle that includes the egg stage, four juvenile stages and the adult stage. A common feature of *Heterorhabditis* and *Steinernema* nematodes is the free-living third stage juvenile (J3), called the dauer juvenile or infective juvenile because it is the infective stage that harbors the mutualistic *Photorhabdus* or *Xenorhabdus* bacteria (Ciche 2007). The infection process starts when the infective juveniles identify and enter the hemocoel (body cavity) of a suitable insect host. Entomopathogenic nematodes invade insects through natural openings, such as the mouth, anus or spiracles, or by disrupting the cuticle using a dorsal tooth in the anterior region of their head, or by secreting enzymes that promote tissue breakdown (Dowds and Peters 2002). Once in the insect hemolymph, nematodes in the infective juvenile stage expel their mutualistic bacterial cells through the mouth by regurgitation (Ciche and Ensign 2003). The infective juveniles recover or exit from the developmentally arrested third, non-feeding stage, triggered by either bacterial or insect food signals (Strauch and Ehlers 1998). The insect hemolymph provides a rich environment for the bacteria that multiply rapidly and the insect dies from septicemia, usually within one or two days (Han and Ehlers 2000). The nematodes feed on the increasing number of bacteria, growing and molting to the fourth stage and then to the adult stage, eventually completing 1–3 generations in the dead insect. Nematodes continue to develop until the food resources of the insect carcass are depleted and as a consequence nematode development is suppressed and infective juveniles accumulate. These non-feeding infective stage larvae exit the insect cadaver and disperse into the soil in search of new hosts. The infective juveniles do not feed in the soil, but they may survive for several months in the absence of a suitable host (Hominick 1990).

Although *Heterorhabditis* and *Steinernema* nematodes have a similar life cycle, they differ in their reproductive strategies. *Heterorhabditis* species have infective juveniles in their first generation that turn into egg-laying hermaphrodites. This means that a single infective juvenile is sufficient to enter the insect host to produce progeny by self-fertilization. In contrast, most *Steinernema* species are dioecious; the infective juveniles develop to become sexually dimorphic adults that reproduce through cross-fertilization or amphimixis. Their progeny either develop to become infective juveniles or adults of the F1 generation (Griffin 2012). Another difference is that *Heterorhabditis* infective juveniles use the cruiser (or foraging) strategy to move actively through the soil in search of a suitable insect host to infect. This tactic involves the response of the nematode cruisers during foraging search to volatile cues emanating from the host or its immediate environment. In contrast, *Steinernema* infective juveniles employ the ambusher (or sit-and-wait) strategy to attack potential hosts that cross their way (Lewis et al. 2006). During this foraging behavior, the nematode ambushers scan during long pauses in a nictation posture (Campbell and Gaugler 1993).

Bacterial colonization of the new infective juveniles (IJ's) at the end of the reproduction cycles is regulated by molecular and cellular interactions between the nematodes and their associated bacteria. The mutualistic bacteria *Xenorhabdus* are localized in a modified ventricular part of the intestine in *Steinernema*, while *Photorhabdus* bacteria are found throughout the intestinal lumen in *Heterorhabditis* (Ciche et al. 2008; Forst and Clarke 2002). The relationship between the mutualistic partners is highly specific and nematodes will only maintain mutualistic associations with their cognate bacteria or very closely related strains (Ffrench-Constant et al. 2003). The nematode–bacteria relationship benefits both participants; the nematodes feed on their associated bacteria and they cannot reproduce in the insect host without the presence of their mutualistic microbes, while the bacteria cannot enter the host hemoceol and cause infection without their nematode vector (Ciche et al. 2006). Although the nematodes may contribute to the death of the infected insect by suppressing the immune system or through the production of virulence factors, *Photorhabdus* and *Xenorhabdus* bacteria are extremely virulent due to the secretion of several toxins and degradative enzymes thereby killing the insect rapidly and providing rich food supply for the developing nematode (Ffrench-Constant et al. 2007a, b; Rodou et al. 2010; Simon et al. 2014). Also, the bacteria create suitable conditions for their nematode host in the dead insect by producing antimicrobial molecules and a large variety of secondary metabolites that inhibit the growth of other microbes, mainly from the insect intestinal microflora, and thus prevent the colonization of the insect cadaver by unwanted microorganisms (Bode 2009).

The identification of novel compounds from nematode–bacteria complexes shows considerable potential for the development of antimicrobial products and other bioactive molecules that may be beneficial to humans. Also, application of entomopathogenic nematodes in an integrated approach in combination with certain insecticides or transgenic plants expressing insecticidal proteins can function in synergy to substantially increase their efficacy in controlling insect pests in the field (Ehlers 2001).

3 Insect Innate Immune Responses

Insects are constantly exposed to microbes in their environment and this requires efficient defense mechanisms to promote their survival and propagation. Insects have developed an innate immune system that allows them to respond rapidly to challenges by different infectious agents (Gillespie et al. 1997). The first line of defense against pathogens involves physical barriers such as the adult exoskeleton, the larval cuticle, the chitinous lining of the tracheae and the peritrophic membrane of the midgut (Davis and Engström 2012) (Fig. 1). Once microbial intruders overcome the external defenses and gain access to the hemocoel, they are recognized as foreign by the insect immune system (Stokes et al. 2015). Immune

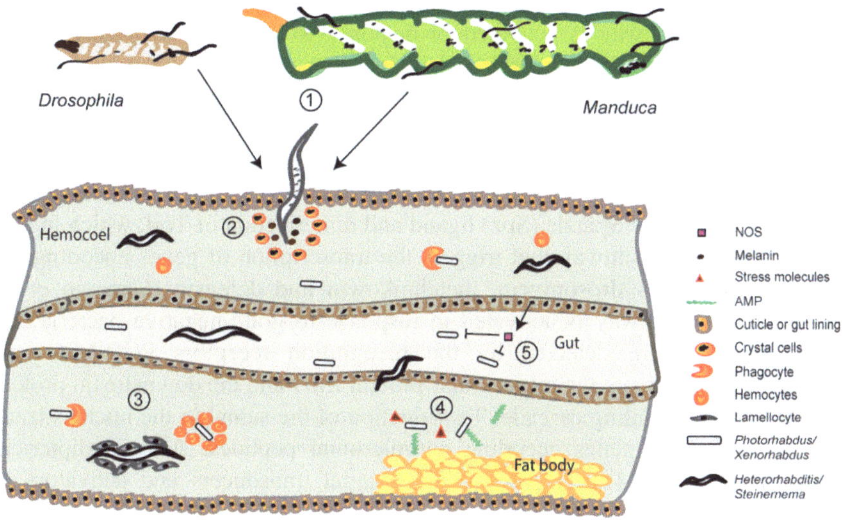

Fig. 1 Insect immune responses against entomopathogenic nematodes and their mutualistic bacteria. *1 Drosophila melanogaster* or *Manduca sexta* larvae restrict the entry of the nematode parasites *Heterorhabditis bacteriophora* or *Steinernema carpocapsae* carrying their bacterial partners *Photorhabdus luminescens* or *Xenorhabdus nematophila*, respectively, by physical barriers consisting of the larval cuticle and the chitinous lining of the tracheae. *2* Once the nematode parasites gain access into the insect hemocoel, crystal cells migrate towards the site of entry to release melanin through the activation of the prophenoloxidase cascade. Melanin deposition promotes wound healing and prevents the pathogens from dissemination. *3* Immune recognition of the invading pathogens (nematodes and their associated bacteria) triggers the cellular arm of the insect innate immune response. Cellular immune reactions include phagocytosis that is directed against *P. luminescens* and *X. nematophila*, nodulation that involves hemocyte aggregation around the bacteria, and encapsulation by lamellocytes in response to the nematode parasites. *4* Activation of the humoral immune response leads to the synthesis and secretion of antimicrobial peptides (AMP) and other stress-related molecules from the fat body into the hemolymph. *5* The presence of the pathogens in the insect gut induces the expression of nitrogen oxide synthase (NOS) from the gut epithelia, which leads to the synthesis of nitrogen oxide that prevents *P. luminescens* bacteria from crossing the gut wall

recognition leads to the induction of downstream signaling pathways that in turn result in the production of effector molecules and the activation of immune responses against the microbial invaders (Buchon et al. 2014). The insect immune system consists of humoral and cellular responses, which involve a wide variety of molecules and defense mechanisms. These include the synthesis and secretion of antimicrobial peptides and proteins into the hemolymph, the production of reactive oxygen and nitrogen species, proteolytic cascades that lead to hemolymph clotting and melanization, and circulating leukocyte-like immune cells (termed hemocytes) that regulate cellular defense mechanisms (Jiang et al. 2010; Lemaitre and Hoffmann 2007; Wojda 2016).

The insect innate immune system is controlled by signaling pathways that are partially conserved in vertebrate animals. The two most investigated pathways, especially in the fruit fly *Drosophila melanogaster*, are the Nuclear Factor κB (NF-kB) pathways Toll and Immune Deficiency (IMD) (Hetru and Hoffmann 2009). The Toll pathway is activated in response to gram-positive bacteria and fungi (Lindsay and Wasserman 2014). Toll is a transmembrane receptor that was first found to be involved in development of the dorso-ventral axis in *Drosophila* embryos (Moussian and Roth 2005). Microbes or microbe-derived compounds are recognized by the proteins upstream of Toll, such as the circulating peptidoglycan recognition proteins PGRP-SA and PGRP-SD (Kurata 2014). Microbial recognition leads to cleavage of the Spätzle (Spz) ligand and dimerization of Toll, which in turn activates a signaling pathway that triggers the transcription of genes encoding the antimicrobial peptides drosomycin, metchnikowin and defensin (Ganesan et al. 2011). The IMD pathway is activated in response to gram-negative bacteria and viruses. Microbes are detected by the recognition receptors PGRP-LC and PGRP-LE, which activate the intracellular protein IMD and the downstream protein complexes of the signaling cascade. Transduction of the signal to the nucleus leads to the expression of genes encoding antimicrobial peptides, such as diptericin (Myllymäki et al. 2014). The Janus kinase/signal transducers and activators of transcription (JAK/STAT) pathway plays a role in *Drosophila* embryogenesis and in regulating the expression of immune molecules, such as the Turandot stress factor proteins and of thioester-containing proteins (TEP), which share sequence homology with vertebrate complement factor C3 and act as opsonins (Myllymäki and Rämet 2014). JAK/STAT signaling also participates in regulating the proliferation of hemocytes in response to tissue damage and tumor development (Amoyel et al. 2014). The c-Jun N-terminal kinase (JNK) pathway is induced by stress stimuli and upon wounding, and participates in wound healing processes. It is also implicated in hemocyte activation, encapsulation and regulation of antimicrobial peptide gene expression (Rämet et al. 2002; Ríos-Barrera and Riesgo-Escovar 2013).

The insect humoral immune response is mostly mediated by the fat body, a diffuse organ that is the largest organ in the hemocoel and analogous to the mammalian liver (Fig. 1). Other organs that also produce humoral factors include epithelial tissues and hemocytes (Kounatidis and Ligoxygakis 2012). Humoral effector molecules mainly involve antimicrobial peptides that are synthesized by the activation of immune signaling pathways and act against fungi, bacteria, viruses and parasitoids (Pinheiro and Ellar 2006; Mylonakis et al. 2016). Antimicrobial peptides are small cationic proteins that are non-detectable in naïve insects, but are rapidly produced, secreted and accumulated in the hemolymph at high concentrations in response to microbial infection. Most antimicrobial peptides are amphiphilic molecules that can specifically disrupt the bacteria membrane by forming a pore. They can also be classified according to their structures, some have intramolecular disulfide bonds, while some are linear peptides (Rolff and Schmid-Hempel 2016; Wiesner and Vilcinskas 2010). There are seven classes of antimicrobial peptides identified in *Drosophila*: Diptericin, Attacin, Drosocin and Cecropin are directed against gram-negative bacteria; Defensin acts against gram-positive bacteria; Drosomycin and Metchnikowin have anti-fungal properties (Imler and Bulet 2005). Interestingly, Attacin, Drosocin, Cecropin and Defensin are also found in other insects; whereas Drosomycin, Diptericin and Metchnikowin are only found in fruit flies (Lemaitre and Hoffmann 2007). Immune effectors identified in lepidopteran insects (moths and butterflies) include C-type lectins, hemolin, serpins, hemolymph proteases (Casanova-Torres and Goodrich-Blair 2013).

The insect cellular immune response involves the action of hemocytes, which participate in the regulation of cellular immune mechanisms, such as phagocytosis, encapsulation and nodulation (Lavine and Strand 2002) (Fig. 1). Hemocytes are grouped according to their morphological features into three types: plasmatocytes, lamellocytes and crystal cells (Ribeiro and Brehélin 2006). Plasmatocytes consist of the major type of hemocytes and perform the function of phagocytosis, a complex process that involves the cellular internalization and degradation of bacteria and other foreign particles or dead cells (Gold and Brückner 2015). Phagocytosis requires a large variety of receptors, and the engulfed particles are subsequently destroyed within phagosomes (Ulvila et al. 2011). Lamellocytes are responsible for the encapsulation of large foreign bodies, such as parasitoid eggs or nematode parasites, which cannot be phagocytosed by the smaller plasmatocytes (Vlisidou and Wood 2015). The encapsulation response involves recognition of the foreign object, recruitment and attachment of lamellocytes that surround it, and simultaneous activation of the insect hematopoietic organ (lymph gland) that rapidly increases the proliferation of lamellocytes in the hemocoel (Honti et al. 2014). In turn, this leads to the development of large nodules consisting of aggregates of hemocytes and adhered bacterial cells. This immune reaction is known as nodulation or nodule formation (Satyavathi et al. 2014). Crystal cells are responsible for the synthesis of enzymes, such as prophenoloxidase, which participate in the melanization process (Crozatier and Meister 2007; Williams 2007).

The melanization response in insects involves the deposition of melanin to invading microorganisms. This mechanism requires the oxidation of phenols to quinones, which are then polymerized to become melanin (Eleftherianos and Revenis 2011; Lu et al. 2014). The melanization reaction needs phenoloxidase, an enzyme that is stored in hemocytes, which catalyzes the oxidation of mono- and diphenols to orthoquinones. The activation of the pro-enzyme prophenoloxidase to active phenoloxidase requires protease cascades triggered by the detection of specific microbial patterns, such as lipopolysaccharides, lipoteichoic acid, peptidoglycans, and β-1,3-glucan, by host immune cell surface recognition receptors (Cerenius et al. 2008; Tang 2009). Certain serine proteases and serpins have been identified to regulate the activation of melanization in insects (Gulley et al. 2013; Veillard et al. 2016). The activity of phenoloxidase is thus used to detect the activation of the melanization immune response. At the wound site, clotting takes place to prevent the loss of hemolymph and promote wound healing. It can also assist in recruiting hemocytes to the site of injury to suppress the spread of pathogens in the insect body (Theopold et al. 2014).

4 Insect Immune Response to Mutualistic Bacteria

Investigations into the immune response of insects against the entomopathogenic bacteria *Photorhabdus* and *Xenorhabdus* have mainly used the lepidopteran caterpillar models *Manduca sexta* and *Galleria mellonella* together with the adult fruit fly *D. melanogaster* model (Castillo et al. 2011). In the studies involving *M. sexta*, caterpillars were injected with bacteria directly, thus circumventing the need for *H. bacteriophora* infective juvenile nematodes. While work on *M. sexta* does not benefit from the availability of genetics and genomics tools, the robust nature of the late instar *M. sexta* larva enables the direct injection of double-stranded RNA (dsRNA) straight into the insect hemocoel for gene silencing through systemic RNAi (Reynolds and Eleftherianos 2008).

To determine whether *P. luminescens* and *P. asymbiotica* bacteria are recognized by the insect immune system, the expression of three major pattern recognition protein genes *Hemolin*, *Immulectin-2* and *Peptidoglycan Recognition Protein 1A* (*PGRP-1A*) in *M. sexta* was examined (Eleftherianos et al. 2006a, b). It was shown that expression of all three recognition genes in *M. sexta* fat body is strongly induced by infection with the pathogens, but they are expressed at only very low levels in unchallenged insects. It was also demonstrated that RNAi-mediated knockdown of any one of these genes results in increased sensitivity to *Photorhabdus* infection. Interestingly, silencing of *Immulectin-2* was found to prevent normal activation of phenoloxidase, which was associated with the reduced ability of the larvae to encapsulate the bacteria. These results revealed that *Photorhabdus* bacteria are recognized by the *M. sexta* immune system and that the induced expression of certain recognition protein genes is an important component of the host immune defense against this pathogen.

Following initial detection of *Photorhabdus* bacteria by host pattern recognition proteins, certain antibacterial peptides (i.e. Attacin, Cecropin, Lebocin, Lysozyme, Moricin) are over-expressed in *M. sexta* fat body. This implies that the antibacterial response of the insect host not only is deployed but also exerts an important (although ultimately ineffective) defense against infection by this pathogen (Eleftherianos et al. 2008). The *M. sexta* immune system can be efficiently primed by prior infection with non-pathogenic bacteria so that immunity to infection by

Photorhabdus is subsequently enhanced (Eleftherianos et al. 2006a, b). Using systemic RNAi to prevent the upregulation of individual immune genes in *M. sexta* revealed that the effect of silencing any of the pattern recognition protein genes is greater than silencing any of the antimicrobial peptide genes. This is because knocking down the expression of single pattern recognition proteins prevents the expression of downstream antimicrobial peptides in hemolymph leading to decreased *P. luminescens* growth, which slows down insect death. Interestingly, older *M. sexta* larvae are less able to induce the transcription of recognition and antimicrobial genes in response to *P. luminescens* suggesting a critical role for host developmental stage in the immune response against this pathogen (Eleftherianos et al. 2008).

Following release of *Photorhabdus* into the insect hemolymph by the infective juveniles, the first response of the host immune system is to phagocytose or encapsulate the invading bacteria (Eleftherianos et al. 2010a, b). *P. luminescens* infection of *M. sexta* larvae causes the appearance of hemocytes with an extreme spreading ability that may play a role in nodule formation (Dean et al. 2004a, b). The presence of these hyper-spreading cells in response to *P. luminescens* is not a pathological effect of infection but a discrete reaction of the insect immune system. It was further shown that the precursor of the plasmatocyte-spreading peptide, a cytokine involved in the hemocyte spreading process, is expressed in the *M. sexta* fat body, but not hemocytes, in response to *P. luminescens* infection, and that RNAi-mediated knockdown of this precursor in both fat body and hemolymph plasma significantly increases insect susceptibility to the pathogen and leads to an overall reduction in cellular immune capacity (Eleftherianos et al. 2009a, b).

One of the most highly expressed genes in *P. luminescens* is *prtA,* which encodes an RTX (repeats in toxin)-like metalloprotease (Bowen et al. 2003; Cabral et al. 2004; Marokhazi et al. 2007; Valens et al. 2002), which targets a small number of *M. sexta* hemolymph proteins that include Serine Protease Homolog 3 (SPH3) (Felföldi et al. 2009). Indeed, silencing the *P. luminescens*-induced expression of *SPH3* in *M. sexta* fat body and hemocytes using systemic RNAi results in higher insect sensitivity to the pathogen (Felföldi et al. 2011). Inhibition of *SPH3* expression has no effect on the transcription of certain pathogen recognition genes, but strongly represses the transcription of multiple antimicrobial effector genes including the gene encoding prophenoloxidase. The impaired ability of the *SPH3* knockdown insects to defend themselves against *P. luminescens* infection is therefore due to interference with effectors that are expressed downstream of SPH3. These findings strongly suggest the importance of SPH3 in the insect innate immune system because it controls the induction of a large number of effector molecules. This makes its destruction an ideal strategy for the highly virulent pathogenic bacterium *P. luminescens.* Also, it has been shown that *P. luminescens*-induced expression of nitric oxide synthase (NOS) in the gut of *M. sexta* following oral infection regulates the immune response to the pathogen through the production of NO (Eleftherianos et al. 2009a, b) (Fig. 1). RNAi-mediated silencing of *NOS* in *M. sexta* larvae orally infected by *P. luminescens* substantially reduces insect survival. This is due to decreased levels of NO in the gut wall of NOS-deficient insects, which promotes the rapid migration of the pathogen from the gut into the hemolymph. The above findings indicate that according to the location of infection by *Photorhabdus* bacteria, different insect tissues and immune molecules expressed in these tissues perform distinct functions to defend the host against the pathogen.

In the greater wax moth (*G. mellonella*) model, it has been shown that injection of an insecticidal toxin produced by the pathogenic bacteria *Xenorhabdus ehlersii* into *G. mellonella* induces the expression of a large number of hemolymph proteins in the infected larvae and upregulates the transcription of immune-related genes in the fat body, which confirms the strong interaction of entomopathogenic bacteria with the insect humoral immune response (Shi et al. 2013). Another study has further shown that *P. luminescens* bacteria co-injected with host-derived extracellular nucleic acids (DNA or RNA) into *G. mellonella* larvae are less pathogenic compared to the pathogens injected alone (Altincicek et al. 2008). The protective effect is partly explained by the activation of host humoral and cellular immune responses, as shown by the sharp increase in antibacterial peptide activity in the hemolymph and the higher numbers of hemocytes in the hemocoel. Proteomic analysis of hemolymph samples from larvae inoculated with nucleic acids has identified several immune proteins that may enhance the host response to *P. luminescens* challenge. Regarding cellular immune processes against entomopathogenic bacteria in *G. mellonella*, infection of *P. asymbiotica* bacteria can boost NO synthesis in the hemocytes of last instar larvae and sessile pharate prepupae of this insect, an effect that is abolished following treatment with a NOS inhibitor (Krishnan et al. 2006). Interestingly, changes in NO production levels are irrelevant to the activation of the prophenoloxidase cascade, and co-injection of the pathogen with the NOS inhibitor does not promote the survival of larvae and pupae. These results denote that overproduction of NO by *P. asymbiotica* in *G. mellonella* hemocytes acts to a certain extent as a cytotoxic factor towards this host. However, a later study reported the identification and purification of an intracellular protein from *Xenorhabdus budapestensis* bacteria, which causes severe melanization and death through injection of *G. mellonella* larvae due to the rapid overactivation of the prophenoloxidase cascade (Yang et al. 2012). *Photorhabdus* bacteria have also been studied for their priming effects on *G. mellonella* larvae. Injection of heat-killed *P. luminescens* into *G. mellonella* provides a dose- and time-dependent protection to a subsequent infection with the bacteria as well to infection with *H. bacteriophora* or *S. carpocapsae* nematodes (Wu et al. 2014). The protective effect is related to changes in hemocyte density (especially an increase in plasmatocyte numbers), stimulation of hemocyte encapsulation, phagocytic ability and antibacterial peptide activity. A similar work proved that lipopolysaccharide (LPS) extracted from *P. luminescens* is responsible for the activation of the host immune system that results in the enhanced immune protection in *G. mellonella* larvae (Wu et al. 2015a). In contrast, the priming effect is not observed in *G. mellonella* pre-injected with a sub-lethal dose of the purified toxin PirA2B2, because prior treatment with this virulence factor diminishes cellular immune activity (hemocyte numbers, phagocytosis and encapsulation) in the larvae (Wu et al. 2015b).

In the model insect *D. melanogaster*, novel information on the immune signaling pathways and defense reactions that are activated in response to *Photorhabdus* and *Xenorhabdus* bacteria has recently started to emerge. Some of these studies have analyzed the molecular basis of the immune response of the fly against entomopathogenic nematode–bacteria complex and the conclusions are discussed in Sect. 6. It has been previously reported that *D. melanogaster* adult flies, as well as *M. sexta* and *G. mellonella* larvae, accumulate black pigments upon injection of the purified metalloprotease PrtS that derives from *P. luminescens* culture supernatants (Held et al. 2007). The authors use fly genetics to show that black pigmentation is due to the activation of the melanization cascade because injection of recombinant PrtS in the *D. melanogaster* melanization mutant *Black Cells* fails to develop dark pigments. Interestingly, wild-type and melanization mutant flies show increased levels of mortality to PrtS injection, implying that death is not due to the melanization response. The fly immune response to entomopathogenic bacteria can also be regulated by preinfection with non-pathogenic bacteria and the presence of endosymbiotic bacteria. A previous investigation on the regulation of immune signaling by entomopathogens in *D. melanogaster* has shown that injection of *P. luminescens* or *X. nematophila* is lethal to wild-type flies and induces the transcription of antimicrobial peptide genes (Aymeric et al. 2010). However, prior challenge with non-pathogenic *Escherichia coli* prolongs the survival of wild-type flies but not IMD mutants, suggesting that the protective effect is probably IMD-dependent. In addition, it was lately shown that although the co-occurrence of both *Wolbachia* and *Spiroplasma* endosymbionts in adult *D. melanogaster* does not affect the survival of the flies to *P. luminescens* infection and the replication of the pathogen in the endosymbiont-carrying flies, it results in the transcriptional regulation of several immune genes that are controlled by the Toll, IMD, JAK/STAT and JNK signaling pathways (Shokal et al. 2016). Finally, nitrite levels in certain wild-type *D. melanogaster* strains have been found to be particularly low upon infection with *P. luminescens* and *NOS* mutant flies exhibit increased sensitivity to the bacteria compared to controls, indicating that NO is an important component of the fly immune response to this entomopathogen (Eleftherianos et al. 2014).

Other studies in non-model insect species have also started to identify host immune system components with a potential role in opposing *Photorhabdus* or *Xenorhabdus* infection. A previous study has shown that changes in intracellular cyclic adenosine monophosphate levels in larval hemocytes of the forest tent caterpillar moth *Malacosoma disstria* increases the clearance of *X. nematophila* from the infected insects (Gulii et al. 2009). Another work has examined the effect of bacterial culture supernatants from *P. luminescens* and *P. temperata* and found differential gene expression in treated adult sweetpotato whiteflies, *Bemisia tabaci* (Shrestha and Lee 2012). The investigators found upregulation of genes associated with immunity and nervous system functions (*knottin* and *acetylcholine receptor*, *acetylcholine esterase* and *sodium channel*, respectively), downregulation of genes participating in metabolism (*cytochrome p450* and *carboxylesterase*), but no changes in the transcription of genes with a role in development (*ecdysone receptor*), reproduction (*vitellogenin*) and stress (*heat-shock proteins 70 and 90*, and *small heat-shock protein*). A more recent study analyzed the transcriptional profile in hemocytes and fat body of cotton bollworm *Helicoverpa armigera* larvae challenged with heat-killed *P. luminescens* and found several differentially regulated genes that take part in immunity, detoxification, development and metabolism (Zhao et al. 2013). Among the differentially regulated immune-related genes, there were several pattern recognition receptors, immune signaling molecules, antimicrobial peptides, various immune proteins, detoxication proteins and cellular immunity-related proteins. These results suggest that immune priming of *H. armigera* leads to strong transcriptional modulation of immune genes that may affect the outcome of a subsequent pathogenic infection.

5 Immune Evasion by Mutualistic Bacteria

Just as mammalian pathogens have developed a number of strategies for evading the adaptive immune system (Baxt et al. 2013), entomopathogenic bacteria have also evolved various means of avoiding or subverting different aspects of the insect innate immune system. *Photorhabdus* and *Xenorhabdus* bacteria use one or more subversion tactics to defeat immune defenses, including the antibacterial peptide response, cellular reactions, and the prophenoloxidase/melanization system (Nielsen-LeRoux et al. 2012). For example, *P. luminescens* PrtA toxin has been shown to target a large number of *M. sexta* hemolymph proteins encoding molecules involved in immune recognition, immune signaling and regulation, and regulation of the coagulation cascade (Felföldi et al. 2009).

X. nematophila bacteria are pathogenic to the beet armyworm *Spodoptera exigua*. Injection of dead bacteria into larvae induces the transcription of several antibacterial peptide genes, including *cecropin*, whereas challenge with live pathogens fails to upregulate transcript levels of *cecropin* (Ji and Kim 2004). Inhibition of antibacterial peptide activity is not associated with the eicosanoid pathway because treatment with arachidonic acid does not revert the inhibitory effect of live *X. nematophila* and treatment with dexamethasone does not reduce the transcriptional induction of *cecropin* by dead bacteria. Similar results were also obtained from a recent study in the red palm weevil *Rhynchophorus ferrugineus* confirming the antibacterial peptide inhibitory activity of live, but not dead, *X. nematophila* endosymbionts (Binda-Rossetti et al. 2016). Interestingly, the CS03 strain of *Xenorhabdus bovienii* proved highly sensitive to the insect antibacterial peptides cecropin A, cecropin B, colistin and polymyxin B, and unable to inhibit the phenoloxidase response and as a result these bacteria are not virulent upon injection into *Spodoptera littoralis* and *G. mellonella* larvae (Bisch et al. 2015). Genetic manipulation of *X. nematophila* showed that the leucine-responsive regulatory protein (Lrp), which regulates mutualism and pathogenicity, is also responsible for interfering with the expression of antibacterial peptides in *M. sexta*, because in contrast to wild-type bacteria, *lrp* mutants are not able to suppress *cecropin A* gene transcription and they are sequestered into nodules (Cowles et al.

2007). As a result, *X. nematophila lrp* mutants have reduced virulence towards *M. sexta* larvae. Also, *X. nematophila* subpopulations can reduce or lose their virulence ability over time and this is accompanied by the inability to suppress antimicrobial peptide gene transcript levels as well as hemocyte aggregation and nodule formation in *M. sexta* (Park et al. 2007). The opacity gene *opaB* has been proposed to be associated with the defects in *X. nematophila* virulence and host immunosuppression.

The effect of entomopathogenic bacteria on cellular immune processes has long been described and recently robust methods have been developed to identify and cluster known and novel bacterial effectors that participate in the manipulation of host cellular defenses (Dowling and Hodgson 2014). *X. nematophila* bacteria have been shown to efficiently infect and replicate in *D. melanogaster* cultured cells as well as in artificial media (Owuama 2003). During in vitro growth, *X. nematophila* produce two separate cytolytic functions against two distinct insect cell types (Brillard et al. 2001). The first takes place when the bacteria reach the stationary phase and causes the formation of vacuoles in hemocytes and necrosis of the cells, mainly granulocytes. The second hemolytic activity occurs during the stationary phase of the bacterial growth and causes shrinkage and necrosis of plasmatocytes. This effect is abolished when the *flhD* gene that resides in the flagellar master operon *flhDC* of *X. nematophila* is mutated, which decreases the virulence of the pathogen towards *S. littoralis* larvae. In the *D. melanogaster* model, an elegant work involving time-lapse confocal microscopy showed that infection of fly embryos with the pathogen *P. asymbiotica* causes a 'freezing' phenotype of the phagocytically competent embryonic hemocytes, which renders the phagocytes incapable of engulfing the bacteria (Vlisidou et al. 2009). The 'freezing' effect is probably produced by the bacterial toxin makes caterpillars floppy (Mcf) that is encoded by the *mcf1* gene (see also the Chapter by Andrea Dowling in this volume). The Mcf toxin has been shown previously to cause apoptosis of *M. sexta* hemocytes and form a major virulence factor of *Photorhabdus* pathogens (Daborn et al. 2002). Interestingly, hemocytes in *D. melanogaster* embryos mutant for the small GTPase Rac do not develop the 'freezing' phenotype upon injection with this toxin. In contrast, the facultative intracellular pathogen *P. asymbiotica* uses the two-component KdpD/KdpE sensor kinase/response regulator to avoid phagocytosis and persist for several days within phagocytes of *M. sexta* larvae, which ultimately results in insect death (Vlisidou et al. 2010). This pathogen can also persist in cultured mammalian macrophages and in vitro the KdpD/KdpE system is upregulated in response to insect hemocytes, but not to hemolymph plasma. This is a strategy used by certain virulent pathogens to survive inside human neutrophils (O'Loughlin et al. 2010). Another *Photorhabdus* species, *P. temperata*, has been recently shown to produce the compound benzaldehyde that is toxic to *G. mellonella* larvae in a dose-dependent manner and is a potent inhibitor of the insect cellular immune system because it strongly suppresses the formation of melanotic nodules together with phenoloxidase activity in the hemolymph, which accelerates

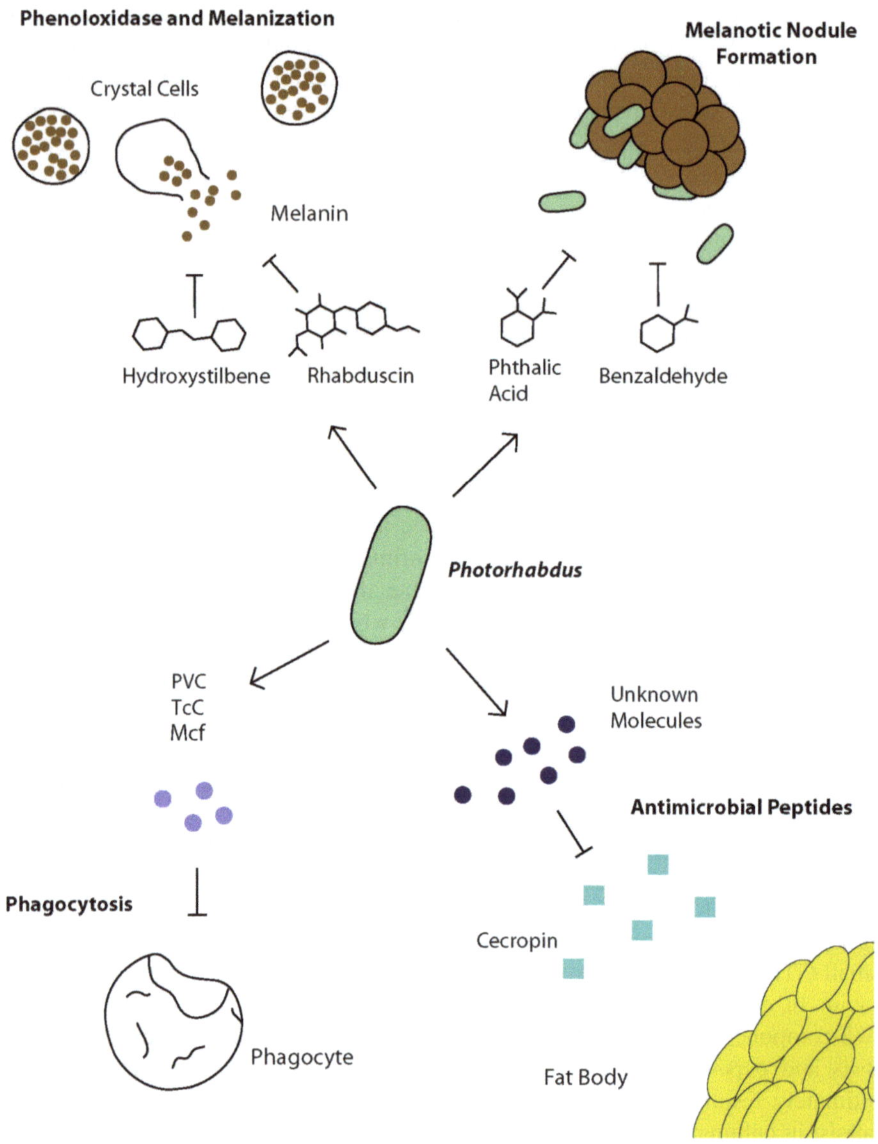
Phenoloxidase and Melanization
Crystal Cells
Melanin
Hydroxystilbene
Rhabduscin
Melanotic Nodule Formation
Phthalic Acid
Benzaldehyde
Photorhabdus
PVC
TcC
Mcf
Phagocytosis
Phagocyte
Unknown Molecules
Antimicrobial Peptides
Cecropin
Fat Body

◄ **Fig. 2** Immune evasion strategies of the pathogen *Photorhabdus luninescens*. The products of the entomopathogen *P. luminescens* are shown in their interactions with the insect immune system as inhibiting phagocytosis, the melanization response, and the production of antimicrobial peptides. In particular, the *P. luminescens* factors *Photorhabdus* virulence cassettes (PVC), toxin complex C (TcC), and makes caterpillars floppy (Mcf) (*bottom left)* inhibit insect phagocytes through modifications of the actin cytoskeleton, with Mcf-based modifications specifically producing a "freezing" phenotype marked by a distinct immobilization of the phagocytes (Fig. 2). Under the phenoloxidase and melanization heading (*top*), hydroxystilbene and rhabduscin are shown to inhibit the melanization response, as mediated by the rupturing of crystal cells, while phthalic acid and benzaldehyde are shown to additionally inhibit the formation of melanotic nodules that can serve to limit the spread of bacteria within the insect. The production of antimicrobial peptides by the insect fat body (*bottom left*) is also addressed by the virulence factors of *P. luminescens* through the production of molecules that degrade cecropin, though these molecules have yet to be identified (Fig. 2)

insect death (Ullah et al. 2015). In the cutworm *S. littoralis* and the locust *Locusta migratoria*, the nodulation reaction is directly linked to the presence of a functional type three secretion system (TTSS) in *P. luminescens*, as TTSS mutants elicit much fewer nodules compared to the wild-type bacteria (Brugirard-Ricaud et al. 2005).

A major host immune interference strategy of entomopathognic bacteria is the subversion of the phagocytosis function (Vallet-Gely et al. 2008). Inhibition of phospholipase A_2 (PLA_2), the major enzyme of eicosanoid biosynthesis, has been previously reported as a common function between *Photorhabdus* and *Xenorhabdus* for inducing host immune suppression through the disruption of cellular reactions (Kim et al. 2005; Park et al. 2004, 2005). In particular, it is known that live *X. nematophila* bacteria disrupt phagocytosis of granular cells and plasmatocytes in *S. exigua* larvae by synthesizing and secreting inhibitors of PLA_2 (Shrestha and Kim 2007). The authors found that treatment with the PLA_2 inhibitor dexamethasone as well as with the metabolite benzylideneacetone decreases hemocyte phagocytosis, whereas treatment with arachidonic acid, the precursor that is enzymatically metabolized to eicosanoids, rescues the phagocytic activity of hemocytes. Live and dead *P. luminescens and P. temperata* bacteria have been shown to be phagocytosed by *M. sexta* hemocytes at much lower levels compared to non-pathogenic *E. coli* bacteria (Au et al. 2004). In addition, culture supernatants from *P. luminescens*, but not *P. temperata*, grown in vitro contain secreted molecules that circumvent phagocytosis of *E. coli* cells. Alternatively, phagocytosis and other cellular immune activities are prevented by these pathogens by targeting hemocytes and reducing hemocyte viability and in the case of *P. luminescens* by producing molecules that cause severe morphological changes to the phagocytes. *P. luminescens* also uses the Tc toxin complexes, which are also found in the human pathogen *Yersinia*, to inhibit phagocytosis of insect cells (Lang et al. 2010). The TccC3 and TccC5 components act together to cause actin clustering; the first by inducing actin polymerization and the second by activating signal pathways that promote the formation of stress fibers.

Photorhabdus and *Xenorhabdus* bacteria are able to prevent the activation of the prophenoloxidase cascade or even suppress the activity of phenoloxidase mature enzyme in the hemolymph of the insect host (Fig. 2). In a previous study, activation of the prophenoloxidase system was linked to PLA_2 inhibition by *X. nematophila* in *S. exigua* (Park and Kim 2003). It was shown that PLA_2 inhibitors also suppress prophenoloxidase activation in the hemolymph as well as the hemolymph plasma and the inhibitory factor can be isolated from the stationary phase culture broth of *X. nematophila*. This suggests that the bacteria secrete molecules that counteract the melanization cascade in *S. exigua* larvae. A similar result was also obtained for the PLA_2 inhibitor, the monoterpenoid compound benzylideneacetone, which also acts as suppressor of activated phenoloxidase in the hemolymph of *Plutella xylostella* larvae, without altering phenoloxidase gene expression in the hemocytes (Song et al. 2011). *P. luminescens* bacteria achieve inhibition of activated phenoloxidase in the hemolymph of *M. sexta* larvae through the secretion of a hydroxystilbene antibiotic compound that is synthesized both in vivo and in vitro (Eleftherianos et al. 2007). Mutating the gene *stlA* encoding the precursor for the production of the stilbene results in reduced ability of the pathogen to inhibit phenoloxidase together with reduced pathogenicity and persistence of the bacteria in the host. Interestingly, RNAi-mediated silencing of the *prophenoloxidase* gene in *M. sexta* does not affect larval survival upon infection with wild-type *P. luminescens* or *stlA* mutant bacteria, indicating a direct relationship between phenoloxidase inhibition by this compound and its effect on pathogen virulence. Another study also identified additional genes with inhibitory properties in *Photorhabdus* (Eleftherianos et al. 2009a, b). Bacterial supernatants or broths from *P. luminescens* or *P. asymbiotica* cosmid clones in *E. coli* possess strong inhibitory activity towards *M. sexta* hemolymph phenoloxidase as well as increased pathogenicity, persistence in the host and reduced nodule formation. The identified genes are mapped to a locus homologous to the *maltodextrin phosphorylase* locus in *E. coli* and other gram-negative bacteria. More recently, an aromatic compound identified as phthalic acid was isolated from *P. temperata* cultures and shown to be toxic to *G. mellonella* larvae and to inhibit phenoloxidase activity and nodule formation, and to possess antibacterial activity (Ullah et al. 2014). Also, *P. luminescens* and *X. nematophila* bacteria synthesize the rhabduscin, a tyrosine derived amidoglycosyl- and vinyl-isonitrile product, a potent inhibitor of phenoloxidase in *G. mellonella* hemolymph, which was found to be localized at the outer part of *X. nematophila* cells and *E. coli* cells expressing this molecule (Crawford et al. 2012). This finding is important because it shows that certain entomopathogenic bacteria have developed elegant ways to fight host immunity and therefore increase their virulence capacity. Another study involving RNA-seq, proteomics and phenotype microarrays in *P. asymbiotica* has confirmed the inhibitory ability of rhabduscin by demonstrating that the heterologously expressed and purified aglycon precursor molecule of rhabduscin acts as inhibitor of the alternative mammalian complement pathway (Mulley et al. 2015).

6 Immune Defense Against Entomopathogenic Nematodes

An important part of the life cycle of entomopathogenic nematodes is their initial interaction with the non-self innate antimicrobial immune defense of the infected insect. During the initial stages of infection, entomopathogenic nematodes may elicit a rapid cellular and humoral immune response in the insect host that could lead to the encapsulation, melanization and killing of the parasites.

Previously it has been reported that larvae of the Japanese beetle *Popillia japonica* display a potent encapsulation and melanization response against the entomopathogenic nematodes *H. bacteriophora*, *S. carpocapsae*, and *S. scapterisci*, but not against the *S. glaseri* nematodes. Interestingly, adults of the house cricket *Acheta domesticus* exhibit a strong immune response against *H. bacteriophora*, *S. carpocapsae*, and *S. glaseri*, but not against the *S. scapterisci* nematodes (Wang et al. 1994). Given that in the wild, *S. glaseri* nematodes are associated with scarab larvae and *S. scapterisci* nematodes usually infect mole crickets, it is possible that the insect host immune response is more potent against entomopathogenic nematodes that are not natural pathogens of the specific insect species. Another piece of evidence showing that the insect cellular immune response against entomopathogenic nematodes varies substantially among insect species is that the in vivo encapsulation of *H. marelatus* nematodes occurs in a very small number of prepupae from the Colorado potato beetle *Leptinotarsa decemlineata* (Armer et al. 2004). The authors succeeded in isolating a heat-labile factor from the hemolymph of the infected insects, which causes the mutualistic bacteria *P. luminescens* to switch from their primary to their secondary form, but this unknown compound is inactive against the nematodes. A similar study further revealed that the insect host immune response toward entomopathogenic nematodes is also specific not only to the nematode species, but also to the strain of the parasite (Li et al. 2007). In particular, it was shown that *S. glaseri* nematodes are more virulent to *P. japonica* larvae compared to *H. bacteriophora*, which is due to the reduced encapsulation and melanization response to *S. glaseri* in these insects. However, the authors found that *S. glaseri* nematodes are not able to develop and reproduce at optimum levels in *A. domesticus*, as previously reported (Wang et al. 1994), and this could be partly attributed to variation in the infectivity properties of the nematode strains used in these two studies. A more recent study has confirmed the substantial variation in the insect immune capacity to entomopathogenic nematode infections (Ebrahimi et al. 2011). It was demonstrated that *H. bacteriophora* and *Steinernema feltiae* nematodes are encapsulated at much higher levels in *L. decemlineata* prepupae than in *G. mellonella* larvae and the nematode encapsulation response is completed significantly earlier in the former than in the latter insect host; therefore it was concluded that *L. decemlineata* is a more responsive host to nematode infection than *G. mellonella*. Curiously, *S. carpocapsae* infective juveniles are not encapsulated in larvae of the coleopteran *Rhynchophorus ferrugineus*, which fails to activate any other type of immune responses against the worms, and nematode parasitism is accompanied by a sharp decrease in the number of circulating hemocytes in the infected insects (Manachini et al. 2013).

In more recent years, the incorporation of the *Drosophila* model in the anti-nematode immunity field has started to reveal the number and nature of genes that control the activation of host defense mechanisms in response to entomopathogenic nematodes. Although *Drosophila* is not a natural host for *Heterorhabditis* or *Steinernema*, recent studies have established protocols for infecting *Drosophila* flies and larvae with entomopathogenic nematodes in the presence or absence of their associated *Photorhabdus* and *Xenorhabdus* bacteria, respectively (Castillo et al. 2012; Dobes et al. 2012; Yadav et al. 2015). These studies have shown that nematode-bacteria complexes are potent pathogens of *Drosophila* and that certain entomopathogenic nematodes lacking their mutualistic bacteria are still pathogenic to flies and larvae (Castillo et al. 2012; Yadav et al. 2015). The use of entomopathogenic nematodes with *Drosophila* has emerged as a tractable system for investigating the genetics of nematode parasitism and elucidating the function of molecules that promote nematode persistence in the infected insect.

A recent study used next generation RNA-sequencing to analyze the transcriptional profile of adult flies infected by symbiotic or axenic *H. bacteriophora* nematodes (containing or lacking their *P. luminescens* bacteria, respectively), or *P. luminescens* bacteria alone (Castillo et al. 2015). Bioinformatic analysis of the data shows that infection with *P. luminescens* or *H. bacteriophora* produces distinct transcriptomic profiles in the fly and that the transcriptome of flies infected by symbiotic nematodes is a combination of the transcriptomes from flies infected by axenic nematodes and their mutualistic bacteria, separately. In particular, *P. luminescens* infection induces several genes encoding recognition and antibacterial effector molecules as well as genes that are involved in translational repression and response to stress. Interestingly, *H. bacteriophora* infection regulates a large number of genes that are involved in lipid homeostasis and metabolism, neural function and nociception, stress response, and genes with potential function in nematode recognition and anti-nematode properties. Both pathogens strongly modulate the transcription of known immune-related genes that control immune anti-pathogen processes in the fly.

Another genome-wide microarray analysis of the transcriptional response of *D. melanogaster* larvae to *H. bacteriophora* nematodes carrying their *P. luminescens* bacteria has identified a large number of genes that are differentially regulated upon infection with the worms (Arefin et al. 2014). More specifically genes that are significantly upregulated include several known immune genes and genes with putative immune properties, while genes that are markedly downregulated are mainly grouped into the oocyte maturation pathway, the Wnt signaling pathway and the ubiquitin-mediated pathway. Interestingly, the authors reported genes encoding certain thioester-containing proteins, clotting factors and cuticular molecules that are also differentially regulated in *D. melanogaster* upon infection with the symbiotic worms.

Further work has demonstrated the power of using *Drosophila* for studying the molecular/genetic basis of insect immune responses against infections by entomopathogenic nematodes. Infection of *D. melanogaster* larvae with *H. bacteriophora* symbiotic worms results in the transcriptional activation of four antimicrobial peptide genes (Hallem et al. 2007). The antimicrobial peptide response is specific to *P. luminescens* bacteria harbored in the nematodes because axenic worms fail to induce the transcription of antimicrobial peptide genes. It was also shown that *H. bacteriophora* nematodes induce the transcription of several immune-related genes in adult *D. melanogaster*, but injection of *P. luminescens* bacteria alone results in lower levels of gene transcription in the flies (Castillo et al. 2013). Inactivation of *D. melanogaster* transglutaminase, a conserved component of clotting cascades in insects and humans, results in decreased aggregation of zymosan beads and increased sensitivity of larvae to *H. bacteriophora* infection (Wang et al. 2010). Two clotting factors (gp150 and fondue) have also been shown to participate in the *D. melanogaster* anti-nematode immune response (Hyrsl et al. 2011), and a homolog of thioester-containing complement protein 3, a basement membrane component (glutactin), a recognition protein (GNBP-like 3) and several small peptides have been found to contribute to the control of *H. bacteriophora* infection in *D. melanogaster* larvae (Arefin et al. 2014). In addition, it has been shown that *S. carpocapsae* symbiotic nematodes are more virulent to *D. melanogaster* larvae than *H. bacteriophora*, and they can upregulate the expression of certain antimicrobial peptide genes as well as induce the melanization pathway, the activation of which is suppressed by mutualistic *X. nematophila* bacteria (Peña et al. 2014). More recently, it was reported that the chitinase-like protein imaginal disc growth factor 3 (IDGF3) plays a regulatory role in the epithelial immune response of *D. melanogaster* larvae against infection with *H. bacteriophora* nematodes by controlling the activation of signaling pathways associated with wound healing (Kucerova et al. 2016). *Idgf3* loss-of-function mutant larvae show high sensitivity to nematode infection, which is accompanied by defects in hemolymph clotting and wound healing processes.

Results generated in the *Drosophila* model provide valuable information on the regulation of host genes that are activated or repressed following infection with the two pathogens, the nematodes and their associated bacteria, either separately or together. Similar studies will provide exciting clues on the molecular immune events that may take place in insects upon infection with potent entomopathogenic nematode–bacteria complexes. Such large-scale transcriptomic analyses set the stage for functional studies aimed at identifying the exact role of key factors in the *Drosophila* anti-nematode and antibacterial immune response.

7 Immune Evasion by Entomopathogenic Nematodes

Once the entomopathogenic nematodes gain access to the hemolymph, they must face the insect humoral and cellular response. In order to release their mutualistic bacteria and complete their life cycle in the insect hemocoel, the nematodes must evade or counteract the host immune system. Evasion of immune defenses by entomopathogenic nematodes can be accomplished by three possible strategies: (i) the nematodes are recognized as self and therefore prevent the induction of immune responses from the host, (ii) the nematodes avoid detection by the host immune system and therefore they are not recognized as non-self, and (iii) the nematodes employ mechanisms to suppresses the host immune response after being recognized as non-self (Dunphy and Webster 1986, 1987). Immune evasion strategies used by entomopathogenic nematodes involve their migration to remote tissues where they cannot be detected by the host immune system (anatomical seclusion), the secretion of molecules or removal of proteins from the hemolymph and their sequestration by the nematode cuticle that prevents immune detection (camouflage or mimicry), and the production of molecules that results in the active suppression of host immune functions (interference) (Brivio et al. 2005; Castillo et al. 2011).

Mechanisms for inhibiting the insect host's antibacterial peptide immune response have been reported in *Heterorhabditis* and *Steinernema* nematodes. Production of cecropin, which is one of the main insect antibacterial peptides responsible for removing bacteria from the hemolymph, can be inhibited by a specific protease that is synthesized and secreted by *H. bacteriophora* nematodes as well as by phase 1 variants of their mutualistic *P. luminescens* bacteria (Jarosz 1998) (Fig. 3). The authors demonstrated that this protease is able to suppress Cecropin synthesis in *G. mellonella* larvae as well as in other lepidopteran and hymenopteran insects. The antibacterial immune response in *G. mellonella* can also be prevented by the body surface of the nematode *S. feltiae* (Brivio et al. 2006). It has been shown that injection of purified nematode cuticles in insects challenged with bacteria results in slower removal of the bacteria from the hemocoel, which is accompanied by reduced levels of antimicrobial peptides in the hemolymph of the infected insects. This antibacterial peptide inhibitory effect is associated with the binding of certain hemolymph proteins by *S. feltiae* cuticular lipids (Fig. 3). The investigators nicely demonstrated that pretreatment of nematode cuticles with enzymes that degrade lipids leads to the loss of the inhibitor activity, and by performing rescue experiments they showed that simultaneous injection of purified hemolymph proteins with bacteria into insects that have previously received nematode cuticles leads to the restoration of the antimicrobial activity. Similar experiments with *S. carpocapsae* and *X. nematophila* showed that infection with live nematodes and bacteria can suppress the antibacterial peptide immune response of *R. ferrugineus*, but the inhibitory effect is not present when insects are injected with dead microorganisms (Binda-Rossetti et al. 2016). In addition, injection of

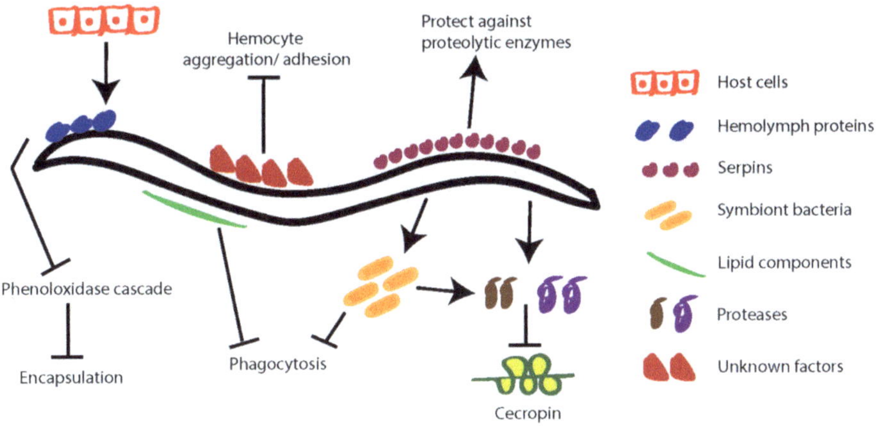

Fig. 3 Immune evasion by entomopathogenic nematodes. *Steinernema carpocapsae* nematodes are able to use their body surface lipid components to sequester certain immunocompetent host hemolymph proteins, which leads to reduced phenoloxidase activity and melanin production. This in turn results in decreased encapsulation ability. Entomopathogenic nematodes use their cuticular lipids or expel their mutualistic bacteria to inhibit the phagocytic function of host hemocytes. Trascriptomic analysis of *S. carpocapsae* nematodes has led to the identification of a serpin that protects the nematode from host proteolytic enzymes and reduces clot formation. *Heterorhabditis bacteriophora* nematodes and their associated bacteria *Photorhabdus luminescens* secrete proteases that suppress the synthesis of the antimicrobial peptide cecropin during infection of the insect host. *S. carpocapsae* nematodes also secrete unknown factors that affect hemocyte numbers and aggregation as well as prevent the adhesion of plasmatocytes and granular cells during infection in lepidopteran insect hosts

isolated nematode cuticles fails to induce the antimicrobial peptide response in the hemolymph of the challenged insects.

Steinernema nematodes have developed tactics to disable the insect immune system in order to create suitable conditions in the host before they expel their mutualistic bacteria. They do this by modulating the immune function of hemocytes, which form the core of the insect cellular immune response. For instance, *G. mellonella* hemocytes are able to recognize *H. bacteriophora* infective juveniles, but not *S. carpocapsae* and *S. glaseri* nematodes (Ebrahimi et al. 2011). Consequently, the insect cellular immune response can differ considerably among nematode species or even strains within the same species. *S. feltiae* nematodes are able to use their cuticle in order to remove certain hemolymph proteins that function as opsonins, and as a result they suppress the phagocytic ability of host hemocytes (Brivio et al. 2010) (Fig. 3). Inhibition of hemocytic function can lead to a weakened host immune system that promotes nematode migration in the host and enhances pathogenicity. The nematodes also interfere with the host defense by producing factors that target the hemocytes in the hemocoel. An in vitro study has found that the *S. carpocapsae-X. nematophila* nematode–bacteria complex produces factors that act against hemocytes from larvae of the lepidopteran *Mythimna unipuncta* (Ribeiro et al. 1999). The secreted factors were separated and shown to possess unsticking and cytotoxic properties towards plasmatocytes and granular cells, respectively. Although not tested, production of those factors in vivo can probably affect the number of hemocytes as well as the aggregation and encapsulation response against the nematodes (Fig. 3). Indeed, another study has shown that third stage juveniles of *S. carpocapsae* nematodes release exudates that prevent the adhesion of plasmatocytes and granular cells during infection of *G. mellonella* and *M. disstria*, and inhibit the removal of cells from preformed hemocyte monolayers in vitro (Walter et al. 2008). This results in reduced phenoloxidase activity in the hemolymph and lower encapsulation of the nematodes. Interestingly, these effects are only associated with live nematodes, which suggests that the inhibitory factors are probably not derived from the nematode cuticle. Elegant work has further identified a chymotrypsin-like serine protease from the mixture of molecules that are produced during the parasitic stage of *S. carpocapsae* nematode (Toubarro et al. 2009). This serine protease was purified and biochemically/molecularly characterized. Results from functional assays have revealed that this serine protease forms a major pathogenicity factor in *S. carpocapsae*, because the purified protease induces early apoptosis of insect cells and causes disruption of the insect gut. Similarly, another trypsin-like serine protease from *S. carpocapsae* was purified and shown to possess potent activity against *G. melonella* hemocytes by blocking hemocyte spreading and causing severe morphological changes to the hemocytes (Balasubramanian et al. 2010). However, experiments involving coincubation of insect hemocytes and *S. feltiae* nematodes revealed that the cuticle of the parasites plays a crucial role in escaping immune recognition by removing hemolymph proteins from the insect host and using them as shield that allows the nematodes to suppress hemocyte aggregation and melanotic encapsulation, and therefore persist in the host (Mastore and Brivio 2008).

Entomopathogenic nematodes can interfere with the insect phenoloxidase response using their cuticle or by secreting molecules that actively suppress the melanization cascade. Challenge with live or dead *S. feltiae* nematodes has been previously shown to decrease the activity of the prophenoloxidase system; this effect is due to reduced protease activity in the hemolymph of the infected insects (Brivio et al. 2002). Injection of purified cuticles into insects or in vitro incubation with insect hemolymph substantially reduces prophenoloxidase activity; however, the inhibitory effect is abolished when nematode cuticles are denatured with SDS. The same research group subsequently reported that the prophenoloxidase inhibitory effect of *S. feltiae* is due to the sequestration of certain immunocompetent hemolymph proteins by lipid compounds on the body surface of the nematodes, which in turn results in reduced *in vitro* melanotic encapsulation of the parasites (Brivio et al. 2004). Molecular and biochemical characterization of a compound produced during the infective stage of *S. carpocapsae* led to the identification of a novel virulence factor, a chymotrypsin-like protease, the purified form of which is able to prevent the in vitro activation of the prophenoloxidase enzyme in hemolymph plasma from *G. mellonella* larvae and suppress the in vivo encapsulation and melanization response against the nematodes (Balasubramanian et al. 2009). Similar studies have also identified a trypsin-like secreted protease from

S. carpocapsae nematodes, which exhibits potent in vitro inhibitory properties towards *G. mellonella* hemolymph prophenoloxidase (Balasubramanian et al. 2010). More recently, a *S. carpocapsae* serine protease inhibitor (serpin) was detected in a transcriptomic study and then purified and expressed (Toubarro et al. 2013a) (Fig. 3). The serpin is highly expressed during the invasive stage and before the nematodes cross the midgut barrier. Functional characterization of the recombinant protein showed that the serpin inhibits the activity of lepidopteran digestive enzymes and provides nematodes with protection against hostile proteolytic enzymes. Although the nematode serpin does not inhibit the activation of prophenoloxidase and clot formation in vitro, it is able to prevent the formation of hardened clots and therefore it suppresses the encapsulation process in *G. mellonella*, it disrupts the insect response to wound sealing, and it is part of a complex that is formed by other hemolymph protein, a trypsin-like protein, apolipophorin and hexamerin. Similarly, an expressed and purified Kunitz-like protease inhibitor identified in *S. carpocapsae* was found to interfere with insect recognition proteins by targeting serine protease homologs from lepidopteran species (Toubarro et al. 2013b). The recombinant inhibitor was shown to prevent the in vitro aggregation of hemocytes and encapsulation of beads in *G. mellonella* hemolymph plasma, but it does not affect the activation of hemocytes, clots, and prophenoloxidase activity, although clotting fibers and phenoloxidase activity were slightly modified.

The model lepidopteran *M. sexta* has been previously used together with three species of *Heterorhabditis* to investigate the insect immune response against nematodes carrying or lacking their mutualistic *Photorhabdus* bacteria (Eleftherianos et al. 2010a, b). It was shown that larval mortality is delayed following infection with axenic nematodes compared to insects infected with nematodes containing *Photorhabdus*, or the bacteria alone; however, the nematodes elicit host immune responses to a lesser extent than the bacteria. Transcription of certain recognition and antibacterial genes in *M. sexta* larvae is lower upon infection with axenic nematodes compared to insects that had received bacteria, either with or without their cognate nematodes. Axenic *Heterorhabditis* also do not elicit the phenoloxidase response and aggregation of hemocytes compared to treatments involving *Photorhabdus*. In contrast, the phagocytic ability of *M. sexta* hemocytes is decreased by both axenic and symbiotic nematodes, but not by *Photorhabdus* alone. These findings indicate that both the nematodes and their bacteria contribute separately to the pathogenic modulation of the host immune response. A recent study in the *G. mellonella* model has demostrated that injection of cuticles from *H. bacteriophora* or *S. carpocapsae* cuticles into the insect hemolymph cause the complete shut down of the entire insect immune system, as shown by the dramatic decrease in all cellular responses (hemocyte numbers, hemocyte aggregation, phagocytosis and encapsulation), humoral responses (production of antimicrobial peptides), and phenoloxidase activity (Yi et al. 2015). Interestingly, eicosanoids have been found to regulate the *G. mellonella* anti-nematode immune response because phospholipase A2 inhibition replicate insect immune suppression, co-injection of phospholipase A2 inhibitors with nematode cuticles also suppress host immunity, but immune suppression is rescued in larvae injected with arachidonic acid, a precursor to eicosanoid biosynthesis. The above studies clearly indicate that entomopathogenic nematodes can express a large variety of molecules, many of which are still unidentified, in order to overcome host immune reactions.

In previous and recent years, the use of transcriptomics in combination with molecular and biochemical techniques has led to the identification of more genes that are potentially involved in parasitic processes of entomopathogenic nematodes (Hao et al. 2008, 2012; Simões et al. 2000). Many of those genes encode proteases or protease-like molecules and their functional characterization has confirmed their mode of action and role as potent pathogenicity factors. Previously identified and characterized virulence factors from *Steinernema* nematodes include two aspartic proteases (Balasubramanian and Simões 2013; Balasubramanian et al. 2012a, b), a pepsin-like aspartic protease (Balasubramanian et al. 2012a), an astacin metallo-protease (Jing et al. 2010), a serine protease (Toubarro et al. 2010), an elastase-like serine protease (Hao et al. 2009), and a surface enolase (Liu et al. 2012). Further characterization of the genetic mechanisms that regulate the interaction between each nematode virulence component with the insect immune system will lead to the determination of novel host anti-nematode immune mechanisms.

8 Perspectives and Future Directions

Insects are excellent models for analyzing the functional, molecular and evolutionary basis of the innate immune response to pathogenic and non-pathogenic organisms. Although important information on the nature of immune responses to bacteria, fungi and viruses has been accumulated over the past several years, recent studies have started to reveal insect immune mechanisms that control infections by entomopathogenic nematode parasites. Furthermore, identification of nematode strategies for evading or avoiding the insect immune response have been used as probes for the characterization of novel insect anti-nematode immune reactions. Future studies using genetic and genomic approaches will undoubtedly help us elucidate the evolutionary conservation of anti-nematode immune responses in vertebrate animals (Castillo et al. 2011).

Novel host immune responses against entomopathogenic nematodes are likely to be identified in model systems in which the genomes of both the host and the pathogen are sequenced and both organisms can be genetically manipulated. Although insects such as *D. melanogaster* are well-established genetic models, genetic tools in the entomopathogenic nematodes *Heterorhabditis* have started to develop (Ciche and Sternberg 2007; Moshayov et al. 2013; Ratnappan et al. 2016). The recently established *Heterorhabditis* infection model in *Drosophila* provides a fascinating system for simultaneous investigations of parasitic nematode infection and anti-nematode immune responses in the host (Hallem et al. 2007). The genomes of both organisms have been sequenced, allowing investigation of important features of pathogenesis and immunity using functional genomics, reverse genetics screens, and systems based approaches. The recent completion of the

H. bacteriophora genome raises interesting questions as to which genes promote nematode parasitism and how these might interact with the insect immune system (Bai et al. 2013). Genes that are upregulated during the infection process offer excellent candidates for potential effector molecules that could participate in host immune evasion or suppression. However, robust molecular tools that permit genetic manipulation are required to study the functional role of candidate molecules in *H. bacteriophora* as well as in other entomopathogenic nematodes.

Successful RNA interference has been reported in *H. bacteriophora*. Two previous studies have shown efficient levels of silencing for several target genes by soaking nematodes in a dsRNA solution (Ciche and Sternberg 2007; Moshayov et al. 2013). Recently, another RNAi silencing protocol was developed in which dsRNA is administered by microinjection into the gonad of the nematodes (Ratnappan et al. 2016). This technique provides a major advantage over RNAi by soaking because the knockdown can be achieved in the progeny. Therefore, infective juveniles can be produced that have decreased levels of infection-associated genes for investigating their role in parasitic processes. In addition, transcriptomic analysis of the *S. carpocapsae* response to incubation in insect hemolymph has identified a large number of genes, some of which might participate in molecular processes that modulate nematode adaptation to the insect immune response (Hao et al. 2010). A similar study in *P. luminescens* used fluorescent reporter constructs to examine the transcriptional regulation of bacterial genes that might be involved in insect infection either by encoding potent virulence factors or by molecules that trick the insect immune system (Münch et al. 2008). Another study involving screening of a recombinant DNA library in a non-pathogenic laboratory strain of *E. coli* carrying individual *P. asymbiotica* cosmids resulted in the identification and phenotypic characterization of putative virulence factors and specialized secretion systems that are probably required to overcome certain host immune defenses and cause disease (Waterfield et al. 2008).

The development of reverse genetic tools in entomopathogenic nematodes together with genome-wide transcriptional profiling of the parasites and their associated bacteria in response to the insect and high-throughput screening of genomic libraries will undoubtedly expose host mechanisms directed against one mutualistic partner only or both pathogens together. Functional analysis of candidate effectors of parasitism will potentially uncover mechanisms by which entomopathogenic nematodes circumvent and cripple the insect immune response, as well as the genetic mechanisms involved in parasitic development following invasion, thereby revealing basic strategies of nematode parasitism and their interrelationship with host immune defenses.

In conclusion, entomopathogenic nematode–bacteria complexes offer great potential for identifying novel pathogen infection strategies that could be used as probes for analyzing the molecular interactions between each mutualistic partner with the insect host. These interactions would likely reveal the signaling pathways that regulate host anti-nematode and antibacterial innate immune responses as well as host molecules with antibacterial/anti-nematode immune properties. Such information is invaluable because it will allow a better understanding of entomopathogenic nematode infections and will potentially result in the development and implementation of improved strategies for the biological control of noxious insect pests.

Acknowledgments The authors wish to thank members of the Eleftherianos laboratory for fruitful discussions. Research in the authors' laboratory is funded by grants from the National Institutes of Health—National Institute of Allergy and Infectious Diseases (1R01AI110675 and 1R21AI109517) and the Columbian College of Arts and Sciences at George Washington University.

References

Altincicek B, Stötzel S, Wygrecka M et al (2008) Host-derived extracellular nucleic acids enhance innate immune responses, induce coagulation, and prolong survival upon infection in insects. J Immunol 181:2705–2712

Amoyel M, Anderson AM, Bach EA (2014) JAK/STAT pathway dysregulation in tumors: a *Drosophila* perspective. Semin Cell Dev Biol 28:96–103

Arefin C, Kucerova L, Dobes P et al (2014) Genome-wide transcriptional analysis of *Drosophila* larvae infected by entomopathogenic nematodes shows involvement of complement, recognition and extracellular matrix proteins. J Innate Immun 6:192–204

Armer CA, Rao S, Berry RE (2004) Insect cellular and chemical limitations to pathogen development: the Colorado potato beetle, the nematode *Heterorhabditis marelatus*, and its symbiotic bacteria. J Invertebr Pathol 87:114–122

Au C, Dean P, Reynolds SE et al (2004) Effect of the insect pathogenic bacterium *Photorhabdus* on insect phagocytes. Cell Microbiol 6:89–95

Aymeric JL, Givaudan A, Duvic B (2010) Imd pathway is involved in the interaction of *Drosophila melanogaster* with the entomopathogenic bacteria, *Xenorhabdus nematophila* and *Photorhabdus luminescens*. Mol Immunol 47:2342–2348

Bai X, Adams BJ, Ciche TA et al (2013) A lover and a fighter: the genome sequence of an entomopathogenic nematode *Heterorhabditis bacteriophora*. PLoS ONE 8:e69618

Balasubramanian N, Simões N (2013) Cloning and molecular analysis of the aspartic protease Sc-ASP110 gene transcript in *Steinernema carpocapsae*. Parasitology 140:1158–1167

Balasubramanian N, Hao YJ, Toubarro D et al (2009) Purification, biochemical and molecular analysis of a chymotrypsin protease with prophenoloxidase suppression activity from the entomopathogenic nematode *Steinernema carpocapsae*. Int J Parasitol 39:975–984

Balasubramanian N, Toubarro D, Simões N (2010) Biochemical study and in vitro insect immune suppression by a trypsin-like secreted protease from the nematode *Steinernema carpocapsae*. Parasite Immunol 32:165–175

Balasubramanian N, Nascimento G, Ferreira R et al (2012a) Pepsin-like aspartic protease (Sc-ASP155) cloning, molecular characterization and gene expression analysis in developmental stages of nematode *Steinernema carpocapsae*. Gene 500:164–171

Balasubramanian N, Toubarro D, Nascimento G et al (2012b) Purification, molecular characterization and gene expression analysis of an aspartic protease (Sc-ASP113) from the nematode *Steinernema carpocapsae* during the parasitic stage. Mol Biochem Parasitol 182:37–44

Baxt LA, Garza-Mayers AC, Goldberg MB (2013) Bacterial subversion of host innate immune pathways. Science 340:697–701

Binda-Rossetti S, Mastore M, Protasoni M et al (2016) Effects of an entomopathogen nematode on the immune response of the insect pest red palm weevil: Focus on the host antimicrobial response. J Invertebr Pathol 133:110–119

Bisch G, Pagès S, McMullen JG 2nd et al (2015) *Xenorhabdus bovienii* CS03, the bacterial symbiont of the entomopathogenic nematode *Steinernema weiseri*, is a non-virulent strain against lepidopteran insects. J Invertebr Pathol 124:15–22
Bode HB (2009) Entomopathogenic bacteria as a source of secondary metabolites. Curr Opin Chem Biol 13:224–230
Boemare N (2002) Biology, taxonomy and systematics of *Photorhabdus* and *Xenorhabdus*. In: Gaugler R (ed) Entomopathogenic nematology. CABI International, Wallingford, pp 35–56
Bowen DJ, Rocheleau TA, Grutzmacher CK et al (2003) Genetic and biochemical characterization of PrtA, an RTX-like metalloprotease from *Photorhabdus*. Microbiology 149:1581–1591
Brillard J, Ribeiro C, Boemare N et al (2001) Two distinct hemolytic activities in *Xenorhabdus nematophila* are active against immunocompetent insect cells. Appl Environ Microbiol 67:2515–2525
Brivio MF, Pagani M, Restelli S (2002) Immune suppression of *Galleria mellonella* (Insecta, Lepidoptera) humoral defenses induced by *Steinernema feltiae* (Nematoda, Rhabditida): involvement of the parasite cuticle. Exp Parasitol 101:149–156
Brivio MF, Mastore M, Moro M (2004) The role of *Steinernema feltiae* body-surface lipids in host-parasite immunological interactions. Mol Biochem Parasitol 135:111–121
Brivio MF, Mastore M, Pagani M (2005) Parasite-host relationship: a lesson from a professional killer. Invertebr Surv J 2:41–53
Brivio MF, Moro M, Mastore M (2006) Down-regulation of antibacterial peptide synthesis in an insect model induced by the body-surface of an entomoparasite (*Steinernema feltiae*). Dev Comp Immunol 30:627–638
Brivio MF, Mastore M, Nappi AJ (2010) A pathogenic parasite interferes with phagocytosis of insect immunocompetent cells. Dev Comp Immunol 34:991–998
Brugirard-Ricaud K, Duchaud E, Givaudan A et al (2005) Site-specific antiphagocytic function of the *Photorhabdus luminescens* type III secretion system during insect colonization. Cell Microbiol 7:363–371
Buchon N, Silverman N, Cherry S (2014) Immunity in *Drosophila melanogaster*–from microbial recognition to whole-organism physiology. Nat Rev Immunol 14:796–810
Cabral CM, Cherqui A, Pereira A et al (2004) Purification and characterization of two distinct metalloproteases secreted by the entomopathogenic bacterium *Photorhabdus* sp. strain Az29. Appl Environ Microbiol 70:3831–3838
Campbell JF, Gaugler R (1993) Nictation behavior and its ecological implications in the host search strategies of entomopathogenic nematodes (*Heterorhabditidae* and *Steinernematidae*). Behavior 126:155–169
Casanova-Torres ÁM, Goodrich-Blair H (2013) Immune signaling and antimicrobial peptide expression in Lepidoptera. Insects 4(3):320–338
Castagnola A, Stock SP (2014) Common virulence factors and tissue targets of entomopathogenic bacteria for biological control of lepidopteran pests. Insects 5:139–166
Castillo JC, Reynolds SE, Eleftherianos I (2011) Insect immune responses to nematode parasites. Trends Parasitol 27:537–547
Castillo JC, Shokal U, Eleftherianos I (2012) A novel method for infecting *Drosophila* adult flies with insect pathogenic nematodes. Virulence 3:339–347
Castillo JC, Shokal U, Eleftherianos I (2013) Immune gene transcription in *Drosophila* adult flies infected by entomopathogenic nematodes and their mutualistic bacteria. J Insect Physiol 59:179–185
Castillo JC, Creasy T, Kumari P et al (2015) *Drosophila* anti-nematode and antibacterial immune regulators revealed by RNA-Seq. BMC Genom 16:519
Cerenius L, Lee BL, Söderhäll K (2008) The proPO-system: pros and cons for its role in invertebrate immunity. Trends Immunol 29:263–271
Ciche T (2007) The biology and genome of *Heterorhabditis bacteriophora*. WormBook 20:1–9
Ciche TA, Ensign JC (2003) For the insect pathogen *Photorhabdus luminescens*, which end of a nematode is out? Appl Environ Microbiol 69:1890–1897

Ciche TA, Sternberg PW (2007) Postembryonic RNAi in *Heterorhabditis bacteriophora*: a nematode insect parasite and host for insect pathogenic symbionts. BMC Dev Biol 7:101

Ciche TA, Darby C, Ehlers RU et al (2006) Dangerous liaisons: the symbiosis of entomopathogenic nematodes and bacteria. Biol Control 38:22–46

Ciche TA, Kim KS, Kaufmann-Daszczuk B, Nguyen KCQ, Hall DH (2008) Cell invasion and matricide during *Photorhabdus luminscens* transmission by *Heterorhabditis bacteriophora* nematodes. Appl Environ Microbiol 74:2275–2287

Cowles KN, Cowles CE, Richards GR et al (2007) The global regulator Lrp contributes to mutualism, pathogenesis and phenotypic variation in the bacterium *Xenorhabdus nematophila*. Cell Microbiol 9:1311–1323

Crawford JM, Portmann C, Zhang X et al (2012) Small molecule perimeter defense in entomopathogenic bacteria. Proc Natl Acad Sci USA 109:10821–10826

Crozatier M, Meister M (2007) *Drosophila* haematopoiesis. Cell Microbiol 9:1117–1126

Daborn PJ, Waterfield N, Silva CP et al (2002) A single *Photorhabdus* gene, makes caterpillars floppy (mcf), allows *Escherichia coli* to persist within and kill insects. Proc Natl Acad Sci USA 99:10742–10747

Davis MM, Engström Y (2012) Immune response in the barrier epithelia: lessons from the fruit fly *Drosophila melanogaster*. J Innate Immun 4:273–283

Dean P, Potter U, Richards EH et al (2004a) Hyperphagocytic haemocytes in *Manduca sexta*. J Insect Physiol 50:1027–1036

Dean P, Richards EH, Edwards JP et al (2004b) Microbial infection causes the appearance of hemocytes with extreme spreading ability in monolayers of the tobacco hornworm *Manduca sexta*. Dev Comp Immunol 28:689–700

Dillman AR, Chaston JM, Adams BJ et al (2012) An entomopathogenic nematode by any other name. PLoS Pathog 8:e1002527

Dobes P, Wang Z, Markus R et al (2012) An improved method for nematode infection assays in *Drosophila* larvae. Fly 6:75–79

Dowds BCA, Peters A (2002) Virulence Mechanisms. In: Gaugler R (ed) Entomopathogenic nematology. CABI International, Wallingford, pp 79–98

Dowling AJ, Hodgson DJ (2014) An unbiased method for clustering bacterial effectors using host cellular phenotypes. Appl Environ Microbiol 80:1185–1196

Dunphy GB, Webster JM (1986) Influence of the Mexican strain of *S. feltiae* and its associated bacterium *Xenorhabdus nematophila* on *Galleria mellonella*. J Parasitol 72:130–135

Dunphy GB, Webster JM (1987) Partially characterized components of the epicuticle of dauer juvenile *Steinernema feltiae* and their influence on the haemocyte activity in *Galleria mellonella*. J Parasitol 73:584–588

Ebrahimi L, Niknam G, Dunphy GB (2011) Hemocyte responses of the Colorado potato beetle, *Leptinotarsa decemlineata*, and the greater wax moth, *Galleria mellonella*, to the entomopathogenic nematodes, *Steinernema feltiae* and *Heterorhabditis bacteriophora*. J Insect Sci 11:75

Ehlers RU (2001) Mass production of entomopathogenic nematodes for plant protection. Appl Microbiol Biotechnol 56:623–633

Eleftherianos IG (2009) Novel antibiotic compounds produced by the insect pathogenic bacterium *Photorhabdus*. Recent Pat Antiinfect Drug Discov 4:81–89

Eleftherianos I, Revenis C (2011) Role and importance of phenoloxidase in insect hemostasis. J Innate Immun 3:28–33

Eleftherianos I, Marokhazi J, Millichap PJ et al (2006a) Prior infection of *Manduca sexta* with non-pathogenic *Escherichia coli* elicits immunity to pathogenic *Photorhabdus luminescens*: roles of immune-related proteins shown by RNA interference. Insect Biochem Mol Biol 36:517–525

Eleftherianos I, Millichap PJ, Ffrench-Constant RH et al (2006b) RNAi suppression of recognition protein mediated immune responses in the tobacco hornworm *Manduca sexta* causes increased susceptibility to the insect pathogen *Photorhabdus*. Dev Comp Immunol 30:1099–1107

Eleftherianos I, Boundy S, Joyce SA et al (2007) An antibiotic produced by an insect-pathogenic bacterium suppresses host defenses through phenoloxidase inhibition. Proc Natl Acad Sci USA 104:2419–2424
Eleftherianos I, Baldwin H, Ffrench-Constant RH et al (2008) Developmental modulation of immunity: changes within the feeding period of the fifth larval stage in the defence reactions of *Manduca sexta* to infection by *Photorhabdus*. J Insect Physiol 54:309–318
Eleftherianos I, Felföldi G, Ffrench-Constant RH et al (2009a) Induced nitric oxide synthesis in the gut of *Manduca sexta* protects against oral infection by the bacterial pathogen *Photorhabdus luminescens*. Insect Mol Biol 18:507–516
Eleftherianos I, Waterfield NR, Bone P et al (2009b) A single locus from the entomopathogenic bacterium *Photorhabdus luminescens* inhibits activated *Manduca sexta* phenoloxidase. FEMS Microbiol Lett 293:170–176
Eleftherianos RH, Ffrench-Constant RH, Clarke DJ et al (2010a) Dissecting the immune response to the entomopathogen *Photorhabdus*. Trends Microbiol 18:552–560
Eleftherianos I, Joyce S, Ffrench-Constant RH et al (2010b) Probing the tri-trophic interaction between insects, nematodes and *Photorhabdus*. Parasitology 137:1695–1706
Eleftherianos I, More K, Spivack S et al (2014) Nitric oxide levels regulate the immune response of *Drosophila melanogaster* reference laboratory strains to bacterial infections. Infect Immun 82:4169–4181
Felföldi G, Marokhazi J, Kepiro M et al (2009) Identificationof natural target proteins indicates functions of a serralysin-type metalloprotease, PrtA, in anti-immune mechanisms. Appl Environ Microbiol 75:3120–3126
Felföldi G, Eleftherianos I, Ffrench-Constant RH et al (2011) A serine proteinase homologue, SPH-3, plays a central role in insect immunity. J Immunol 186:4828–4834
Ffrench-Constant R, Waterfield N, Daborn P et al (2003) *Photorhabdus*: towards a functional genomic analysis of a symbiont and pathogen. FEMS Microbiol Rev 26:433–456
Ffrench-Constant RH, Dowling A, Waterfield NR (2007a) Insecticidal toxins from *Photorhabdus* bacteria and their potential use in agriculture. Toxicon 49:436–451
Ffrench-Constant RH, Eleftherianos I, Reynolds SE (2007b) A nematode symbiont sheds light on invertebrate immunity. Trends Parasitol 23:514–517
Forst S, Clarke D (2002) Bacteria-nematode symbiosis. In: Gaugler R (ed) Entomopathogenic nematology. CABI International, Wallingford, pp 57–77
Forst S, Dowds B, Boemare N et al (1997) *Xenorhabdus* and *Photorhabdus* spp.: bugs that kill bugs. Annu Rev Microbiol 51:47–72
Ganesan S, Aggarwal K, Paquette N et al (2011) NF-κB/Rel proteins and the humoral immune responses of *Drosophila melanogaster*. Curr Top Microbiol Immunol 349:25–60
Gillespie JP, Kanost MR, Trenczek T (1997) Biological mediators of insect immunity.. Annu Rev Entomol 42:611–643
Gold KS, Brückner K (2015) Macrophages and cellular immunity in *Drosophila melanogaster*. Semin Immunol 27:357–368
Goodrich-Blair H, Clarke DJ (2007) Mutualism and pathogenesis in *Xenorhabdus* and *Photorhabdus*: two roads to the same destination. Mol Microbiol 64:260–268
Griffin CT (2012) Perspectives on the behavior of entomopathogenic nematodes from dispersal to reproduction: traits contributing to nematode fitness and biocontrol efficacy. J Nematol 44:177–184
Griffin CT, Downes MJ, Block W (1990) Tests of Antarctic soils for insect parasitic nematodes. Antarctic Sci 2:221–222
Gulii V, Dunphy GB, Mandato CA (2009) Innate hemocyte responses of *Malacosoma disstria* larvae (C. Insecta) to antigens are modulated by intracellular cyclic AMP. Dev Comp Immunol 33:890–900
Gulley MM, Zhang X, Michel K (2013) The roles of serpins in mosquito immunology and physiology. J Insect Physiol 59:138–147
Hallem EA, Rengarajan M, Ciche TA (2007) Nematodes, bacteria, and flies: a tripartite model for nematode parasitism. Curr Biol 17:898–904

Han R, Ehlers RU (2000) Pathogenicity, development, and reproduction of *Heterorhabditis bacteriophora* and *Steinernema carpocapsae* under axenic in vivo conditions. J Invertebr Pathol 75:55–58

Hao YJ, Montiel R, Nascimento G et al (2008) Identification, characterization of functional candidate genes for host-parasite interactions in entomopathogenetic nematode *Steinernema carpocapsae* by suppressive subtractive hybridization. Parasitol Res 103:671–683

Hao YJ, Montiel R, Nascimento G et al (2009) Identification and expression analysis of the Steinernema carpocapsae elastase-like serine protease gene during the parasitic stage. Exp Parasitol 122:51–60

Hao YJ, Montiel R, Abubucker S et al (2010) Transcripts analysis of the entomopathogenic nematode *Steinernema carpocapsae* induced in vitro with insect haemolymph. Mol Biochem Parasitol 169:79–86

Hao YJ, Montiel R, Lucena MA et al (2012) Genetic diversity and comparative analysis of gene expression between *Heterorhabditis bacteriophora* Az29 and Az36 isolates: uncovering candidate genes involved in insect pathogenicity. Exp Parasitol 130:116–125

Held KG, LaRock CN, D'Argenio DA (2007) A metalloprotease secreted by the insect pathogen *Photorhabdus luminescens* induces melanization. Appl Environ Microbiol 73:7622–7628

Hetru C, Hoffmann JA (2009) NF-kappaB in the immune response of *Drosophila*. Cold Spring Harb Perspect Biol 1:a000232

Hillyer JF (2016) Insect immunology and hematopoiesis. Dev Comp Immunol 58:102–118

Hominick WM (1990) Entomopathogenic Rhabditid nematodes and pest control. Parasitol Today 6:148–152

Honti V, Csordás G, Kurucz É et al (2014) The cell-mediated immunity of *Drosophila melanogaster*: hemocyte lineages, immune compartments, microanatomy and regulation. Dev Comp Immunol 42:47–56

Hyrsl P, Dobes P, Wang Z et al (2011) Clotting factors and eicosanoids protect against nematode infections. J Innate Immun 3:65–70

Imler JL, Bulet P (2005) Antimicrobial peptides in *Drosophila*: structures, activities and gene regulation. Chem Immunol Allergy 86:1–21

Jarosz J (1998) Active resistance of entomophagous rhabditid *Heterorhabditis bacteriophora* to insect immunity. Parasitology 117:201–208

Ji D, Kim Y (2004) An entomopathogenic bacterium, *Xenorhabdus nematophila*, inhibits the expression of an antibacterial peptide, cecropin, of the beet armyworm, *Spodoptera exigua*. J Insect Physiol 50:489–496

Jiang H, Vilcinskas A, Kanost MR (2010) Immunity in lepidopteran insects. Adv Exp Med Biol 708:181–204

Jing Y, Toubarro D, Hao Y et al (2010) Cloning, characterisation and heterologous expression of an astacin metalloprotease, Sc-AST, from the entomoparasitic nematode *Steinernema carpocapsae*. Mol Biochem Parasitol 174:101–108

Kenney E, Eleftherianos I (2016) Entomopathogenic and plant pathogenic nematodes as opposing forces in agriculture. Int J Parasitol 46:13–19

Kim Y, Ji D, Cho S et al (2005) Two groups of entomopathogenic bacteria, *Photorhabdus* and *Xenorhabdus*, share an inhibitory action against phospholipase A2 to induce host immunodepression. J Invertebr Pathol 89:258–264

Kounatidis I, Ligoxygakis P (2012) *Drosophila* as a model system to unravel the layers of innate immunity to infection. Open Biol 2:120075

Krishnan N, Hyrsl P, Simek V (2006) Nitric oxide production by hemocytes of larva and pharate prepupa of *Galleria mellonella* in response to bacterial lipopolysaccharide: cytoprotective or cytotoxic? Comp Biochem Physiol C Toxicol Pharmacol 142:103–110

Kucerova L, Broz V, Arefin B et al (2016) The *Drosophila* chitinase-like protein IDGF3 is involved in protection against nematodes and in wound healing. J Innate Immun 8:199–210

Kurata S (2014) Peptidoglycan recognition proteins in *Drosophila* immunity. Dev Comp Immunol 42:36–41

Lang AE, Schmidt G, Schlosser A et al (2010) *Photorhabdus luminescens* toxins ADP-ribosylate actin and RhoA to force actin clustering. Science 327:1139–1142

Lavine MD, Strand MR (2002) Insect hemocytes and their role in immunity. Insect Biochem Mol Biol 32:1295–1309

Lemaitre B, Hoffmann J (2007) The host defense of *Drosophila melanogaster.* Annu Rev Immunol 25:697–743

Lewis EE, Campbell J, Griffin C et al (2006) Behavioral ecology of entomopathogenic nematodes. Biol Control 38:66–79

Li XY, Cowles RS, Cowles EA et al (2007) Relationship between the successful infection by entomopathogenic nematodes and the host immune response. Int J Parasitol 37:365–374

Lindsay SA, Wasserman SA (2014) Conventional and non-conventional *Drosophila* Toll signaling. Dev Comp Immunol 42:16–24

Liu H, Zeng H, Yao Q et al (2012) *Steinernema glaseri* surface enolase: molecular cloning, biological characterization, and role in host immune suppression. Mol Biochem Parasitol 185:89–98

Lu A, Zhang Q, Zhang J et al (2014) Insect prophenoloxidase: the view beyond immunity. Front Physiol 5:252

Manachini B, Schillaci D, Arizza V (2013) Biological responses of *Rhynchophorus ferrugineus* (Coleoptera: Curculionidae) to *Steinernema carpocapsae* (Nematoda: Steinernematidae). J Econ Entomol 106:1582–1589

Marokhazi J, Mihala N, Hudecz F et al (2007) Cleavage site analysis of a serralysin-like protease, PrtA, from an insect pathogen *Photorhabdus luminescens* and development of a highly sensitive and specific substrate. FEBS J 274:1946–1956

Mastore M, Brivio MF (2008) Cuticular surface lipids are responsible for disguise properties of an entomoparasite against host cellular responses. Dev Comp Immunol 32:1050–1062

Moshayov A, Koltai H, Glazer I (2013) Molecular characterisation of the recovery process in the entomopathogenic nematode *Heterorhabditis bacteriophora.* Int J Parasitol 43:843–852

Moussian B, Roth S (2005) Dorsoventral axis formation in the *Drosophila* embryo–shaping and transducing a morphogen gradient. Curr Biol 15:R887–R899

Mulley G, Beeton ML, Wilkinson P et al (2015) From insect to man: *Photorhabdus* sheds light on the emergence of human pathogenicity. PLoS ONE 10:e0144937

Münch A, Stingl L, Jung K et al (2008) *Photorhabdus luminescens* genes induced upon insect infection. BMC Genom 9:229

Myllymäki H, Rämet M (2014) JAK/STAT pathway in *Drosophila* immunity. Scand J Immunol 79:377–385

Myllymäki H, Valanne S, Rämet M (2014) The *Drosophila* imd signaling pathway. J Immunol 192:3455–3462

Mylonakis E, Podsiadlowski L, Muhammed M et al (2016) Diversity, evolution and medical applications of insect antimicrobial peptides. Philos Trans R Soc Lond B Biol Sci 371:20150290

Nielsen-LeRoux C, Gaudriault S, Ramarao N et al (2012) How the insect pathogen bacteria *Bacillus thuringiensis* and *Xenorhabdus/Photorhabdus* occupy their hosts. Curr Opin Microbiol 15:220–231

O'Loughlin JL, Spinner JL, Minnich SA et al (2010) *Yersinia pestis* two-component gene regulatory systems promote survival in human neutrophils. Infect Immun 78:773–782

Owuama CI (2003) Invasion of insect blood tissue by *Xenorhabdus bovienii.* Res Microbiol 154:183–189

Park Y, Kim Y (2003) *Xenorhabdus nematophilus* inhibits p-bromophenacyl bromide (BPB)-sensitive PLA2 of *Spodoptera exigua.* Arch Insect Biochem Physiol 54:134–142

Park Y, Kim Y, Stanley D (2004) The bacterium *Xenorhabdus nematophila* inhibits phospholipases A2 from insect, prokaryote, and vertebrate sources. Naturwissenschaften 91:371–373

Park Y, Aliza AR, Stanley D (2005) A secretory PLA2 associated with tobacco hornworm hemocyte membrane preparations acts in cellular immune reactions. Arch Insect Biochem Physiol 60:105–115

Park Y, Herbert EE, Cowles CE et al (2007) Clonal variation in *Xenorhabdus nematophila* virulence and suppression of *Manduca sexta* immunity. Cell Microbiol 9:645–656

Peña JM, Carrillo MA, Hallem EA (2014) Variation in the susceptibility of *Drosophila* to different entomopathogenic nematodes. Infect Immun 83:1130–1138

Peters A (2013) Application and commercialization of nematodes. Appl Microbiol Biotechnol 97:6181–6188

Pinheiro VB, Ellar DJ (2006) How to kill a mocking bug? Cell Microbiol 8:545–557

Rämet M, Lanot R, Zachary D et al (2002) JNK signaling pathway is required for efficient wound healing in *Drosophila*. Dev Biol 241:145–156

Ratnappan R, Vadnal J, Keaney M et al (2016) RNAi-mediated gene knockdown by microinjection in the model entomopathogenic nematode *Heterorhabditis bacteriophora*. Parasit Vectors 9:160

Reynolds SE, Eleftherianos I (2008) RNAi and the insect immune system. In: Beckage NE (ed) Insect immunology. Academic Press, Amsterdam, pp 295–330

Ribeiro C, Brehélin M (2006) Insect haemocytes: what type of cell is that? J Insect Physiol 52:417–429

Ribeiro C, Duvic B, Oliveira P et al (1999) Insect immunity-effects of factors produced by a nematobacterial complex on immunocompetent cells(1). J Insect Physiol 45:677–685

Ríos-Barrera LD, Riesgo-Escovar JR (2013) Regulating cell morphogenesis: the *Drosophila* Jun N-terminal kinase pathway. Genesis 51:147–162

Rodou A, Ankrah DO, Stathopoulos C (2010) Toxins and secretion systems of *Photorhabdus luminescens*. Toxins 2:1250–1264

Rolff J, Schmid-Hempel P (2016) Perspectives on the evolutionary ecology of arthropod antimicrobial peptides. Philos Trans R Soc Lond B Biol Sci 371:20150297

Satyavathi VV, Minz A, Nagaraju J (2014) Nodulation: an unexplored cellular defense mechanism in insects. Cell Signal 26:1753–1763

Shapiro-llan DI, Han R, Dolinski C (2012) Entomopathogenic nematode production and application technology. J Nematol 44:206–217

Shi H, Zeng H, Yang X (2013) An insecticidal protein from *Xenorhabdus ehlersii* stimulates the innate immune response in *Galleria mellonella*. World J Microbiol Biotechnol 29:1705–1711

Shokal U, Yadav S, Atri J et al (2016) Effects of co-occurring *Wolbachia* and *Spiroplasma* endosymbionts on the *Drosophila* immune response against insect pathogenic and non-pathogenic bacteria. BMC Microbiol 16:16

Shrestha S, Kim Y (2007) An entomopathogenic bacterium, *Xenorhabdus nematophila*, inhibits hemocyte phagocytosis of *Spodoptera exigua* by inhibiting phospholipase A(2). J Invertebr Pathol 96:64–70

Shrestha YK, Lee KY (2012) Oral toxicity of *Photorhabdus* culture media on gene expression of the adult sweetpotato whitefly, *Bemisia tabaci*. J Invertebr Pathol 109:91–96

Simões N, Caldas C, Rosa JS et al (2000) Pathogenicity caused by high virulent and low virulent strains of *Steinernema carpocapsae* to *Galleria mellonella*. J Invertebr Pathol 75:47–54

Simon NC, Aktories K, Barbieri JT (2014) Novel bacterial ADP-ribosylating toxins: structure and function. Nat Rev Microbiol 12:599–611

Song CJ, Seo S, Shrestha S et al (2011) Bacterial metabolites of an entomopathogenic bacterium, *Xenorhabdus nematophila*, inhibit a catalytic activity of phenoloxidase of the diamondback moth, *Plutella xylostella*. J Microbiol Biotechnol 21:317–322

Stokes BA, Yadav S, Shokal U et al (2015) Bacterial and fungal pattern recognition receptors in homologous innate signaling pathways of insects and mammals. Front Microbiol 6:19

Strauch O, Ehlers RU (1998) Food signal production of *Photorhabdus luminescens* inducing the recovery of entomopathogenic nematodes *Heterorhabditis* spp. in liquid culture. Appl Microbiol Biotechnol 50:369–374

Tang H (2009) Regulation and function of the melanization reaction in *Drosophila*. Fly 3:105–111

Theopold U, Krautz R, Dushay MS (2014) The *Drosophila* clotting system and its messages for mammals. Dev Comp Immunol 42:42–46

Toubarro D, Lucena-Robles M, Nascimento G et al (2009) An apoptosis-inducing serine protease secreted by the entomopathogenic nematode *Steinernema carpocapsae*. Int J Parasitol 39:1319–1330

Toubarro D, Lucena-Robles M, Nascimento G et al (2010) Serine protease-mediated host invasion by the parasitic nematode *Steinernema carpocapsae*. J Biol Chem 285:30666–30675

Toubarro D, Avila MM, Hao Y et al (2013a) A serpin released by an entomopathogen impairs clot formation in insect defense system. PLoS ONE 8:e69161

Toubarro D, Avila MM, Montiel R et al (2013b) A pathogenic nematode targets recognition proteins to avoid insect defenses. PLoS ONE 8:e75691

Ullah I, Khan AL, Ali L et al (2014) An insecticidal compound produced by an insect-pathogenic bacterium suppresses host defenses through phenoloxidase inhibition. Molecules 19:20913–20928

Ullah I, Khan AL, Ali L et al (2015) Benzaldehyde as an insecticidal, antimicrobial, and antioxidant compound produced by *Photorhabdus temperata* M1021. J Microbiol 53:127–133

Ulvila J, Vanha-Aho LM, Rämet M (2011) *Drosophila* phagocytosis - still many unknowns under the surface. APMIS 119:651–662

Valens M, Broutelle AC, Lefebvre M et al (2002) A zinc metalloprotease inhibitor, Inh, from the insect pathogen *Photorhabdus luminescens*. Microbiology 148:2427–2437

Vallet-Gely I, Lemaitre B, Boccard F (2008) Bacterial strategies to overcome insect defences. Nat Rev Microbiol 6:302–313

Veillard F, Troxler L, Reichhart JM (2016) *Drosophila melanogaster* clip-domain serine proteases: structure, function and regulation. Biochimie 122:255–269

Vlisidou I, Wood W (2015) *Drosophila* blood cells and their role in immune responses. FEBS J 282:1368–1382

Vlisidou I, Dowling AJ, Evans IR et al (2009) *Drosophila* embryos as model systems for monitoring bacterial infection in real time. PLoS Pathog 5:e1000518

Vlisidou I, Eleftherianos I, Dorus S et al (2010) The KdpD/KdpE two-component system of *Photorhabdus asymbiotica* promotes bacterial survival within *M. sexta* hemocytes. J Invertebr Pathol 105:352–362

Walter TN, Dunphy GB, Mandato CA (2008) *Steinernema carpocapsae* DD136: metabolites limit the non-self adhesion responses of haemocytes of two lepidopteran larvae, *Galleria mellonella* (F. Pyralidae) and *Malacosoma disstria* (F. Lasiocampidae). Exp Parasitol 120:161–174

Wang Y, Gaugler R, Cui L (1994) Variations in Immune Response of *Popillia japonica* and *Acheta domesticus* to *Heterorhabditis bacteriophora* and *Steinernema* Species. J Nematol 26:11–18

Wang Z, Wilhelmsson C, Hyrsl P et al (2010) Pathogen entrapment by transglutaminase—a conserved early innate immune mechanism. PLoS Pathog 6:e1000763

Waterfield NR, Sanchez-Contreras M, Eleftherianos I et al (2008) Rapid Virulence Annotation (RVA): identification of virulence factors using a bacterial genome library and multiple invertebrate hosts. Proc Natl Acad Sci USA 105:15967–15972

Waterfield NR, Ciche T, Clarke D (2009) *Photorhabdus* and a host of hosts. Annu Rev Microbiol 63:557–574

Wiesner J, Vilcinskas A (2010) Antimicrobial peptides: the ancient arm of the human immune system. Virulence 1:440–464

Williams MJ (2007) *Drosophila* hemopoiesis and cellular immunity. J Immunol 178:4711–4716

Wojda I (2016) Immunity of the greater wax moth *Galleria mellonella*. Insect Sci. doi:10.1111/1744-7917.12325

Wu G, Zhao Z, Liu C et al (2014) Priming *Galleria mellonella* (Lepidoptera: Pyralidae) larvae with heat-killed bacterial cells induced an enhanced immune protection against *Photorhabdus luminescens* TT01 and the role of innate immunity in the process. J Econ Entomol 107:559–569

Wu G, Yi Y, Lv Y et al (2015a) The lipopolysaccharide (LPS) of *Photorhabdus luminescens* TT01 can elicit dose- and time-dependent immune priming in *Galleria mellonella* larvae. J Invertebr Pathol 127:63–72

Wu G, Yi Y, Sun J et al (2015b) No evidence for priming response in Galleria mellonella larvae exposed to toxin protein PirA2B2 from *Photorhabdus luminescens* TT01: An association with the inhibition of the host cellular immunity. Vaccine 33:6307–6313

Yadav S, Shokal U, Forst S et al (2015) An improved method for generating axenic entomopathogenic nematodes. BMC Res Notes 8:461

Yang J, Zeng HM, Hf Lin et al (2012) An insecticidal protein from *Xenorhabdus budapestensis* that results in prophenoloxidase activation in the wax moth, *Galleria mellonella*. J Invertebr Pathol 110:60–67

Yi Y, Wu G, Lv J et al (2015) Eicosanoids mediate *Galleria mellonella* immune response to hemocoel injection of entomopathogenic nematode cuticles. Parasitol Res 115:597–608

Zhao Z, Wu G, Wang J et al (2013) Next-generation sequencing-based transcriptome analysis of *Helicoverpa armigera* larvae immune-primed with *Photorhabdus luminescens* TT01. PLoS ONE 8:e80146

Part IV
Photorhabdus Asymbiotica and Human Pathogenicity

Photorhabdus asymbiotica as an Insect and Human Pathogen

Alexia Hapeshi and Nick R. Waterfield

Abstract *Photorhabdus asymbiotica* is a species of bacterium that is pathogenic to humans whilst retaining the ability to infect insect hosts. Currently, there are two recognised subspecies, *P. asymbiotica* subsp. *asymbiotica* and *P. asymbiotica* subsp. *australis* with strains isolated from various locations in the USA, Australia, Thailand, Nepal and Europe. Like other species of *Photorhabdus*, *P. asymbiotica* subsp. *australis* was shown to form a symbiotic relationship with a *Heterorhabditis* nematode. In contrast to most strains of *Photorhabdus luminescens*, *P. asymbiotica* can grow at 37 °C and this is a defining factor in its ability to cause human disease. Insights into other adaptations it has undergone that have enabled host switching to occur have come from whole genome sequencing and transcriptomic studies. *P. asymbiotica* has a smaller genome compared to *P. luminenscens* with a lower diversity of insecticidal toxins. However, it has acquired plasmids and several pathogenicity islands in its genome. These encode genes with similarity to effectors or systems found in other known human pathogens such as *Salmonella* and *Yersinia* and are therefore likely to contribute to human pathogenicity. Of crucial importance to virulence is the fact that *P. asymbiotica* undergoes a large metabolic shift at the human host temperature.

Contents

A. Hapeshi · N.R. Waterfield (✉)
Warwick Medical School, University of Warwick, Coventry, UK
e-mail: N.R.Waterfield@warwick.ac.uk

A. Hapeshi
e-mail: A.Hapeshi@warwick.ac.uk

Current Topics in Microbiology and Immunology (2017) 402:159–178
DOI 10.1007/82_2016_29

Published Online: 03 September 2016

1 Introduction

Photorhabdus asymbiotica was originally described in 1989 after being isolated from a patient with a leg ulcer in the USA. This was followed by a search of the records of the Centres for Disease Control and Prevention that revealed three more previously reported cases (Farmer et al. 1989). There are 15 cases reported in the literature and they involve soft tissue infections that are often multifocal, whilst in five cases there was also an associated bacteraemia (see also the current review of clinical cases by John Gerrard in this volume). All of the reported cases occurred in North America and Australia and were generally associated with outdoor activities in warm weather conditions (Gerrard et al. 2004). The existence of additional isolates from Thailand and Nepal (Gerrard et al. 2011; Thanwisai et al. 2012) indicates a widespread geographical distribution of the species. However, epidemiological information on *P. asymbiotica* is likely to be incomplete since infections are probably misdiagnosed, as commonly used automated bacterial identification systems misidentify this pathogen (Gerrard et al. 2003). The fact that the majority of cases reported originate from neighbouring locations may therefore simply reflect the increased awareness of clinicians in those areas regarding the possibility of infection with this unexpected pathogen (Weissfeld et al. 2005). Alternatively, it could point to the existence of specific localised pools of closely related strains, or their symbiont nematodes, that have acquired adaptations enabling them to infect human hosts. So far there are two recognised subspecies of *P. asymbiotica*, *P. asymbiotica* subsp. *asymbiotica*, found in the USA and *P. asymbiotica* subsp. *australis*, which includes the Australian isolates (Gerrard et al. 2006; Peat et al. 2010) and the isolate from Thailand (Thanwisai et al. 2012).

Since the bacteria were first identified from clinical isolates, it was originally believed they did not form the normal symbiotic relationship with the nematode vector. They were therefore originally described as being ‘a-symbiotic’ as the species name suggests. Eventually however, a nematode belonging to the genus *Heterorhabditis* was found to be associated with the clinical *P. asymbiotica* Kingscliff isolate from Australia (Gerrard et al. 2006). The nematode is most closely related to *H. indica*, but is classified as a distinct species, *H. gerrardi* (Plichta et al. 2009). In addition, certain bacterial strains isolated from soil isolates of Japanese *Heterhorhabditis* nematodes are phylogenetically most related to *P. asymbiotica* species (Kuwata et al. 2008). It should be pointed out, however, that *Heterhorhabditid* nematodes vectoring any North American isolates have not yet been identified.

Nevertheless, *Photorhabdus* species have never been found to occur free-living in the soil, even though at least some can grow in soil under laboratory conditions (Bleakley and Chen 1999). The mechanism of transmission to humans also remains unknown. In the case of infection with *P. asymbiotica* strain Kingscliff a preexisting skin injury was reported (Gerrard et al. 2006). However, in other incidences there was no known prior skin lesion (Gerrard et al. 2011). The most likely hypothesis is that the nematode vector penetrates human skin and delivers the *Photorhabdus* subcutaneously enabling establishment of the infection. Interestingly, it remains a formal, but unproven, possibility that as yet unrecognised mammalian or bird species could provide natural hosts for these particular strains and their associated nematodes.

Importantly, all North American and Australian *P. asymbiotica* isolates tested have the ability to grow at temperatures of 37 °C or above. Interestingly, the strains isolated from Australia are able to tolerate even higher temperatures than those isolated from North America. Amongst all the *P. asymbiotica* genospecies so far examined, the only strains not able to grow at temperatures over 34 °C are two strains isolated from Northern Europe (*P. asymbiotica* strain HIT and *P. asymbiotica* strain JUN) (Mulley et al. 2015). These observations suggest that adaptation of the *P. asymbiotica* 'genospecies' to elevated temperatures may have been influenced by geographical location and climatic conditions. While most strains of *P. luminescens* tested cannot grow at elevated temperatures there are examples of some that can (Fischer-Le Saux et al. 1999). It is not known if these would be capable of causing human infection given the opportunity. However, as no clinical cases have been ascribed to any member of the genus other than *P. asymbiotica*, we suggest it implies certain adaptations are required that are not found in the *P. luminescens* genospecies.

2 Insights from the *P. asymbiotica* Genome

Genomic data is an invaluable source of information enabling the identification of factors that have allowed host switching to occur. Currently, a complete genome of a USA isolate (ATCC 43949) (Wilkinson et al. 2009) and two draft genomes of Australian isolates, *P. asymbiotica* Kingscliff (Wilkinson et al. 2010) and *P. asymbiotica* subsp. *australis* DSM 17609 (our unpublished data) are available. Comparison between these and the genome sequences of *P. luminescens* strains has revealed a smaller genome size for *P. asymbiotica* with a concurrent acquisition of novel elements. The genome of *P. asymbiotica* ATCC 43949 is 5,064,808 bp (Wilkinson et al. 2009) compared to that of *P. luminescens* TTO1 at 5,688,987 bp (Duchaud et al. 2003). The *P. asymbiotica* strains harbour extra-chromosomal elements, with the USA strain carrying one plasmid, pPAU1, which bears some homology to the pMT1 plasmid from *Yersinia pestis*. The presence of the latter is necessary for *Y. pestis* to be able to colonise its flea vector. The Australian *P. asymbiotica* strain also carries a pPAU1 homologue (pPAA1), but in addition also carries a second plasmid element, pPAA3. This bears similarity to pCRY from *Yersinia pestis* 91001 and encodes a Type IV pilus system, which may be involved in either plasmid conjugation or, alternatively as a Type IV toxin secretion system. No plasmids have yet been identified from either *P. luminescens* or *P. temperata* isolates despite an early report of plasmid isolation from *P. luminescens* MU2 (Bondi et al. 1999). Interestingly our unpublished Illumina whole genome sequencing of the HIT and JUN isolates has confirmed that these strains also contain sequences homologous to the pPAU1 plasmid. In the case of pPAJ from *P. asymbiotica* strain JUN, we can also confirm that this exists as a closed circle independent of the chromosome. Interestingly in this pPAJ plasmid an alternative origin of replication has inserted into the normal pPAU1 *repA* gene, although the significance of this, if any, is not known.

Even though the exact function of the *P. asymbiotica* plasmids has not been determined, our previous attempts to cure pPAU1 failed. Furthermore, *P. luminescens* TT01 transformed with pPAU1, genetically tagged with an antibiotic resistance marker, becomes non-viable after only two or three subcultures. This suggests there is genetic cross talk between plasmid and chromosome and that the presence of these plasmids is somehow required for normal growth of *P. asymbiotica*. Furthermore, comparisons of whole genome phylogeny of *P. asymbiotica* isolates with the whole plasmid sequence phylogeny shows the same tree topology suggesting that acquisition of the pPAU1-like plasmid was ancestral to the speciation of *P. asymbiotica* within the genus.

Other differences noted between the genomes of *P. asymbiotica* and *P. luminescens* are in Type III secretion systems (T3SS). T3SS are used by pathogens to introduce their own protein effectors into host cells. *P. luminescens* possesses one T3SS with at least one tightly linked effector, LopT that is homologous to the *Yersinia pestis* YopT. YopT is a serine protease that causes disruption of the host cell cytoskeleton (Shao et al. 2002). As is the case with *Yersinia*, *lopT* is found in an operon with *slcT*, which encodes a putative LopT chaperone, so it is likely that the function of the two proteins in *Photorhabdus* is similar to that of the *Yersinia* homologues. Indeed previous work by Brugirard-Ricaud et al. confirmed the role of LopT in preventing phagocytosis of *P. luminescens* by insect hemocytes (Brugirard-Ricaud et al. 2005). At the corresponding T3SS locus in *P. asymbiotica*, however, LopT is absent and instead the T3SS operon encodes a homologue of ExoU from *Pseudomonas aeruginosa*. ExoU is a known phospholipase (Sato et al. 2003) that is the cause of acute lung injury leading to sepsis (Pankhaniya et al. 2004). Additionally, *P. asymbiotica* appears to have acquired a second T3SS structural operon, related to a system from clinical isolates of *Vibrio parahaemolyticus* (Park et al. 2004). Moreover, at a separate genomic location, and likely exported by the T3SS, *P. asymbiotica* also encodes a homologue of *sopB*. SopB is T3SS effector of *Salmonella enterica* serovar Typhimurium and is involved in the maturation of the *Salmonella*-containing vacuole via recruitment of Rab5 (Mallo et al. 2008). Finally, *P. asymbiotica* contains a homologue of *vopS* from *V. parahaemolyticus*, which is another T3SS effector, which modifies small Rho-family GTPases by AMPylation.

Photorhabdus Virulence Cassettes (PVCs) are discrete operons encoding proteins similar to phage tail and Type 6 Secretion System components. They encode macromolecular phage tail-like structures superficially similar to R-type pyocins (Yang et al. 2006). Unlike pyocins however, which target other bacterial cells, the PVCs appear to be used as a mechanism for delivery of *Photorhabdus* effectors into eukaryotic cells (Yang et al. 2006). Each cassette consists of typically 16 genes that encode the structural components of the PVC 'needle complex', followed by one or more genes encoded at the 3′ end of the operon that are homologues of typical T3SS-like toxin effectors. Indeed transient expression of several example effectors in mammalian cells confirmed profound effects of the cell cytoskeleton (Yang et al. 2006). The genome of *P. luminescens* contains six PVC loci, four of which are arranged in tandem (PVC units 1–4). Different putative effectors are associated with each unique PVC cassette. The genome of *P. asymbiotica* encodes only five PVCs (Fig. 1). In particular, PVC units 1–3 are absent from the genome and only the apparently ancestral PVCunit4 is conserved in the same syntenic locus. Interestingly this PVCunit4 encodes effectors distinct from those found on the *P. luminescens* equivalent. The *P. asymbiotica* PVCunit4 encodes two effectors that are homologous to *gogB* from *Salmonella enterica* serovar Typhimurium and *Escherichia coli* and thus may play a role in the mammalian infection process. In contrast, at the equivalent *P. luminescens* locus there are two genes related to

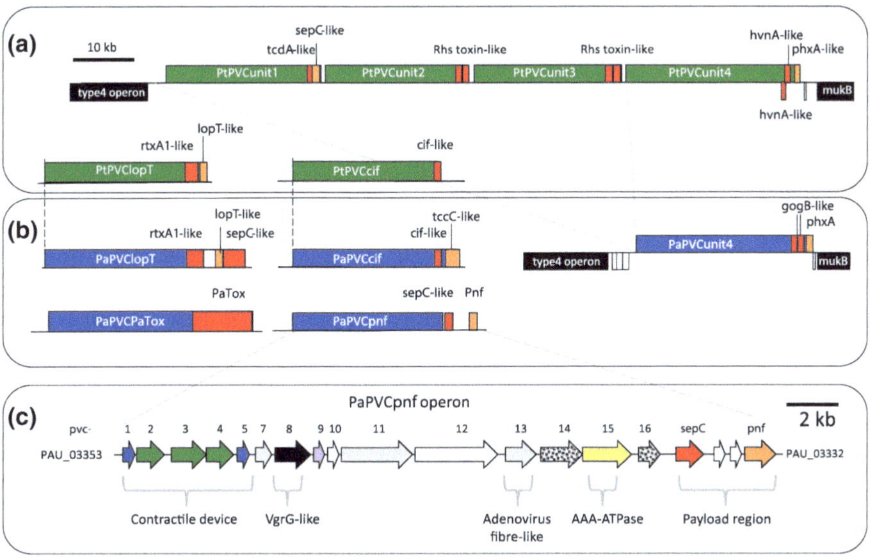

Fig. 1 The *Photorhabdus* virulence cassettes (PVCs). PVCs found in the genome of *P. luminescens* are shown in (**a**) and PVCs found in *P. asymbiotica* 43949 are shown in (**b**). *Dashed lines* between (**a**) and (**b**) show conserved elements, whilst dotted lines indicate the genomic region of *P. luminescens* TT01 that is absent from the corresponding locus of the *P. asymbiotica* 43949 genome. A more detailed schematic representation of the *P. asymbiotica* PVCpnf operon is shown in (**c**)

halovibrin from *Vibrio fischeri*. Halovibrins are secreted proteins that may facilitate the symbiotic relationship between *V. fisheri* and the Bobtail squid (Stabb et al. 2001). As such, it is tempting to speculate that this particular PVC locus might be performing a similar symbiosis role between the bacteria and the *Heterorhabditis* nematode. It should be noted that an analogous, and at least partially homologous, PVC-like 'injectosome' needle complex system is employed by the bacterium *Pseudoalteromonas luteoviolacea* which it uses to control the metamorphosis of the marine worm *Hydroides elegans* (Shikuma et al. 2014). This suggests that the PVC-like elements may be an ancient and a general mechanism for controlling many aspects of interaction with invertebrate hosts, including their development.

Two more PVC elements, PVClopT and PVCcif are conserved between the two species. However, *P. asymbiotica* also contains two additional unique PVC elements: PVCPaTox and PVCpnf. The latter is toxic to insects by injection (and their hemocytes) when heterologously expressed from a cosmid clone in *E. coli* (Yang et al. 2006). The Pnf effector is homologous to the cytotoxic necrotizing factor (CNF1) from *E. coli* (Fabbri et al. 2010). Interestingly, while a PVC operon is present in the equivalent locus in the genome of the Australian strain *P. asymbiotica* Kingscliff, a gene coding for an alternative effector has replaced *pnf*. This highlights the variability and flexibility in these novel delivery systems (Wilkinson et al. 2010). Moreover, the effector of the PVCPaTox element, encoded by gene PAU_02230, has been identified as a *P. asymbiotica* toxin and is termed PaTox (Jank et al. 2013). PaTox has some similarity to the *Salmonella* virulence factor SseI. SseI is an effector of the *Salmonella* pathogenicity island 2 T3SSS that is thought to inhibit migration of immune cells and in particular antigen presenting cells (McLaughlin et al. 2009). The gene encoding PaTox was identified separately following a search in the *P. asymbiotica* genome for the presence of motifs found in the glycosylating toxins of *Clostridium* and *Legionella* (Jank et al. 2013) (see also the Chapter by Andrea Dolwing in this volume).

Another well-known toxin system first discovered in *Photorhabdus* are ABC-type 'Toxin Complexes' encoded by *tc* genes (see also the Chapter by Klaus Aktories in this volume), some of which are responsible for oral toxicity to insects. The toxins are composed of TcA, TcB and TcC subunits. The TcA subunit contains the host cell receptor binding component and forms a homo-oligomeric injection complex which is used to inject the TcB-TcC subunits into the host cell cytoplasm (Gatsogiannis et al. 2013). Different TcC homologues possess different C-terminal domains and at least one of which has been shown to function as an ADP-ribosyltransferase. The genes encoding these were initially seen to be located on four distinct pathogenicity islands in strain *P. luminescens* W14, from which the toxins were first identified and purified (Bowen and Ensign 1998). These pathogenicity islands were named, *tca*, *tcb*, *tcc* and *tcd*. Since genome sequences have become available it is possible to see significant variation in *tc* toxin gene complement and organisation, even between different *P. luminescens* strains. For example, in strain TT01 the *tca* island has undergone deletion of most of *tcaA* and *tcaB* and the *tcb* island has also been mostly deleted. However, a comparison with the *P. asymbiotica* genome reveals even greater differences in *tc* gene complement. For example, in addition to an independent *tcaAB* deletion event, in *P. asymbiotica* an additional *tccC* homologue is encoded on the *tca* island. Furthermore, the *tcd* island is much reduced with four of the equivalent *P. luminescens* genes missing (Wilkinson et al. 2009). These differences are observed in both the American and the Australian sequenced *P. asymbiotica* isolates (Wilkinson et al. 2010). Indeed comparative genomics suggests that the *P. asymbiotica tcd* island is likely to represent a smaller more ancestral state and that the *P. luminescens* strains have acquired additional homologues.

The genome of *P. asymbiotica* also shows a reduced complement of other toxin homologues compared to *P. luminescens.* For example, while *P. luminescens* possesses two homologues of the large Makes Caterpillars Floppy toxin (*mcf1* and *mcf2*), *P. asymbiotica* only carries the *mcf1* homologue. Mcf1 possesses a BH3-like N-terminal domain which induces apoptosis in hemocytes and the insect midgut epithelium (Daborn et al. 2002). Mcf2 on the other hand, has an N-terminal domain that shows similarity to HrmA (Waterfield et al. 2003), a T3SS effector of *Pseudomonas syringae.* Transfection of this N-terminal domain into mammalian NIH3T3 cells resulted in cell death (Waterfield et al. 2003). Expression of either Mcf1 or Mcf2 confers virulence properties to *E. coli* and enables the bacteria to kill insects (Daborn et al. 2002; Waterfield et al. 2003). An additional example can be found with the *pirAB* toxin genes. Unlike *P. luminescens*, which has two loci encoding PirAB toxins, *P. asymbiotica* only encodes one such locus, equivalent to the *plu4093-plu4092* genes of TT01. The PirAB toxins have strong insecticidal activity against Diptera and Lepidoptera (Waterfield et al. 2005; Ahantarig et al. 2009). A crystal structure of the *Vibrio parahaemolyticus* PirA and PirB homologues has revealed similar topology to the *Bacillus* pore-forming Cry toxin, suggesting a similar mode of action (Lee et al. 2015). Finally, *P. asymbiotica* 43949 is missing four haemolysin-encoding loci, which are present in *P. luminescens*. It must be noted, however, that a comparison between the genomes of the Australian *P. asymbiotica* Kingscliff isolate and *P. asymbiotica* 43949 does reveal that the former contains several haemolysin or haemagglutinin genes that are not found in the USA isolate (Wilkinson et al. 2010). Whether these are employed during the infection in the insect or in the mammalian infection process is still unknown.

The differences discussed above in the toxin compliments of *P. asymbiotica* 43949 and *P. luminescens* TT01 are illustrated in a functional genomics comparison between the two species. This follows the results of a screen whereby parts of the two genomes were cloned into cosmids and introduced into *E. coli.* The resulting *E. coli* clones were then tested for toxicity against insects, macrophages, amoebas or nematodes (Waterfield et al. 2008). As can be seen in Fig. 2, a comparison of the genetic regions of the two bacteria, which are capable of conferring toxicity, shows that a number of toxins found in *P. luminescens* TT01 are absent from *P. asymbiotica* 43949, whilst the latter has acquired novel elements. Additionally, there are parts of the genome that are conserved between the two species that do not seem to encode toxic products in *P. luminenscens*, but do so in *P. asymbiotica.* This

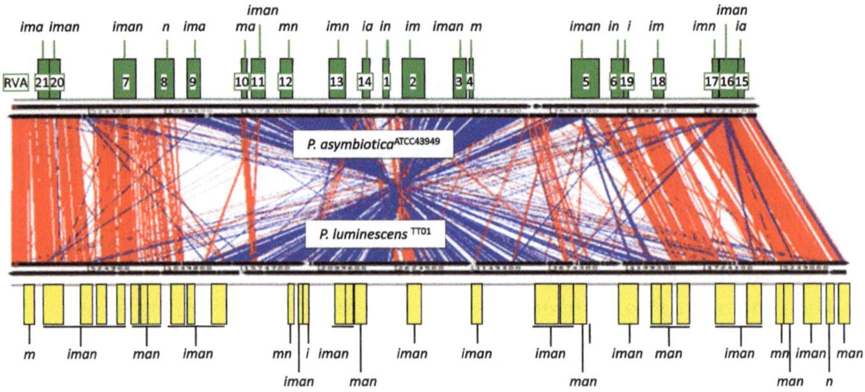

Fig. 2 A comparison of the rapid virulence annotation (*RVA*) functional genomics screens of *P. asymbiotica* 43949 and *P. luminescens* TT01. The image represents an artemis comparison tool alignment of the two genomes. *Boxes above* and *below* the genomes represent regions of the genome containing genes that elaborate toxic products in *E. coli* cosmid clones. *RVA* gain of toxicity screens were performed against *Manduca sexta* insects (*i*), Mammalian macrophages (*m*), *Acanthamoeba polyphaga* amoeba (*a*) and *C. elegans* nematodes (*n*). The *RVA* numbers at the top refer to the relevant regions presented in the supplementary data in (Waterfield et al. 2008), where the *P. asymbiotica RVA* screen was described. The *P. luminescens RVA* data is unpublished

may reflect specific adaptations in *P. asymbiotica* that facilitate infection of the human host.

Genetic differences between *P. asymbiotica* and the other species of the genus involving genes that encode non-toxin products may also be of significance to its life style and its ability to infect humans. One such example is the LPS biosynthesis locus whereby there is a large region that is different between *P. asymbiotica* 43949 and *P. luminescens* TT01. In particular, a 17 kb region of *P. luminescens* TT01 has been replaced by a region of 18 kb that contains putative O-antigen synthesis genes. Elucidation of the structure of the *O*-polysaccharide of various *P. asymbiotica* strains has revealed that there are some subtle differences between Australian and American isolates since there are higher levels of *O*-acetylation in the American strains. However, in general the structure resembles those of other important phylogenetically distant pathogens, such as *Francisella tularensis* or *Shigella dysenteriae* type 7 (Kondakova et al. 2011). Thus, the differences observed between *P. asymbiotica* and *P. luminescens* in this respect may reflect an additional adaptation of the former to its new host.

3 *P. asymbiotica* as an Insect Pathogen

Despite the overall lower diversity of insecticidal toxin genes in the *P. asymbiotica* genome, the ability of the bacterium to infect model insects is unaffected. Indeed, *P. asymbiotica* is more pathogenic against model insect hosts than the other members of the genus. Eleftherianos and coworkers observed that using a dose of approximately 100 bacteria to infect *Manduca sexta* larvae, the LT_{50} of *P. asymbiotica* 43949 is 30 h, whilst that of *P. luminescens* is 56 h (Eleftherianos et al. 2006). Nevertheless, it cannot be ruled out that as a result of the lower diversity of toxin genes in *P. asymbiotica* the bacterium is unable to infect as large a range of insect host species as *P. luminescens*. To our knowledge, the virulence of *P. asymbiotica* against a range of insect hosts has not been tested.

Interestingly, *P. asymbiotica* becomes unable to cause a successful infection in *Manduca sexta* larvae when the infected insects are incubated at 37 °C (Mulley et al. 2015). This suggests that the bacteria are not naturally found in an insect host at this temperature and that they have adapted other mechanisms that are better suited for an infection of mammalian hosts. The apparent increased virulence of *P. asymbiotica* at the lower temperature of 28 °C could be a result of the presence of additional virulence factors that the species has acquired. For example, the PaTox is not found in other *Photorhabdus* species and could contribute to virulence against insects. The toxin consists of a deamidase and a glycosyltransferase domain. The activity of the deamidase domain constitutively activates heterotrimeric G proteins, whilst the glycosyltranferase domain modifies small Rho-family GTPases in their GTP-bound active state. Overall, PaTox activity results in actin cytoskeleton disassembly and thus inhibition of phagocytosis. Pa Tox is active against both insect cells and mammalian cells (Jank et al. 2013). Similarly, the PVCpnf that is specific to *P. asymbiotica* is another potent insect virulence factor. Injection of recombinant PVCpnf into *Galleria melonella* larvae is lethal and hemocytes recovered from the larvae show severe actin cytoskeleton rearrangement (Yang et al. 2006).

It is important to note here that *P. asymbiotica* can be found within insect hemocytes, unlike *P. luminescens* which remains extracellular (Brugirard-Ricaud et al. 2005; Costa et al. 2009; Vlisidou et al. 2010). The rate of uptake of *P. asymbiotica* by hemocytes appears to depend on the subspecies, since Australian isolates seem to be taken up more readily than USA isolates (Wilkinson et al. 2009; Costa et al. 2009). The ability of *P. asymbiotica* to survive in cells means that the pathogen may have alternative uses for factors that are present across the genus. For example, the KdpD/KdpE two component system of *P. asymbiotica* contributes to survival of the bacteria within insect hemocytes (Vlisidou et al. 2010). Vlisidou and coworkers showed that it alone could enable nonpathogenic *E. coli* to resist phagocytic killing by increasing expression of the native high affinity potassium pump Kdp (Vlisidou et al. 2010). How exactly this interferes with the hemocyte ability to kill the phagocytosed bacteria is not clear, but *E. coli* was able to persist long enough to result in the death of injected *Manduca sexta* larvae (Vlisidou et al. 2010). A KdpD/KdpE system is also present in *P. luminescens* but it is not clear whether this would provide an advantage to the bacteria during the infection if they are not found within host cells. Moreover, the ability of *P. asymbiotica* to enter and survive within the host cells implies that toxin molecules synthesised by the bacteria might become released directly into the host cell cytosol without the need for a delivery mechanism. This could allow for example the *P. asymbiotica* Tca BC to act as a functional toxin despite the lack of the pore-forming TcA subunit, as there is no need for transport of the toxin across the host cell membrane. On the other hand, even though *P. asymbiotica* virulence against model insect hosts appears to be undiminished, oral toxicity of the bacteria is attenuated due to the lack of certain *tc* genes, required for full oral toxicity.

4 The Human Host

4.1 Overcoming Host Defences

Many of the immune-defence strategies that *Photorhabdus* must overcome in an insect have parallels in a human host. In fact many aspects of innate immunity at both a mechanistic and even genetic level are conserved between many invertebrates studied and humans (Browne et al. 2013; Buchmann 2014; Kimbrell and Beutler 2001). These could be either part of the humoral response, such as the presence of antimicrobial peptides and production of lysozyme, or the cellular innate immune response, which involves phagocytic attack by hemocytes in the insect or macrophages in the human host. Thus, it may be envisaged that the mechanisms employed by the bacteria to evade the response of the insect can be adapted to address these similar issues encountered in the human host. In support of this notion, it was recently demonstrated that both *P. luminescens* and *P. asymbiotica* can survive and grow in the presence of human serum (Mulley et al. 2015). This suggests that the adaptations that allow *Photorhabdus* to overcome killing by components of the insect serum, provide sufficient protection against human serum. Importantly though, unlike *P. luminescens*, *P. asymbiotica* is also able to survive in the presence of both pig and rabbit serum. Similarly, the aglycon precursor of rhabducin, a known inhibitor of the insect phenol oxidase cascade (Crawford et al. 2012) was also shown to inhibit the human alternative complement pathway (Mulley et al. 2015). Additionally, RNAseq analysis of *P. asymbiotica* revealed the up-regulation of *isnAB* transcription for the synthesis genes of rhabducin at 37 °C, compared to at 28 °C (Mulley et al. 2015). Even though there was no detectable increased secretion of free rhabducin itself (Mulley et al. 2015), most of the molecule is normally found on the cell surface (Crawford et al. 2012). Another *Photorhabdus* secondary metabolite whose levels actually increase at 37 °C and thus may be relevant in the human infection process is iso-propyl-stilbene (IPS). IPS is produced by both *P. luminescens* and *P. asymbiotica* and is an antibiotic compound with multiple other activities that is also required for symbiosis (Joyce et al. 2008; Li et al. 1995) (see also the Chapters by David Clarke and Helge Bode on secondary metabolism also in this volume). Interestingly, IPS also seems to have immunomodulatory properties (Zang et al. 2016) so it could be another example of a factor present in the *Photorhabdus* genome that is being repurposed by *P. asymbiotica* during a human infection. Finally, the secreted metalloprotease PrtA, which is found in both clinical and nonclinical strains of *Photorhabdus* might be involved in defence against the immune system in the human host. This is supported by the identification of several PrtA-sensitive proteins with immune related function, following exposure of insect haemolymph to PrtA digestion in vitro (Felföldi et al. 2009). It remains to be confirmed whether these proteins are PrtA targets in vivo.

Bacterial pathogens can either remain outside host cells or survive and replicate inside them. Furthermore, some that are facultative intracellular pathogens can switch between both strategies. Experiments in vitro using both murine and human macrophage-like cells have shown the *P. asymbiotica* can enter these phagocytic cells (Wilkinson et al. 2009; Costa et al. 2009). Costa et al. (2009), observed that *P. asymbiotica* could actually survive and replicate inside human macrophage-like THP-1 cells. It is tempting to speculate that the ability to survive intracellularly is at least in part mediated by the *P. asymbiotica* SopB homologue discussed previously. Additionally, Australian but not American isolates are able to invade the non-phagocytic HeLa cells (Costa et al. 2009). It is possible that the increased ability of the Australian isolates to invade host cells is due to the presence of the Type 4 secretion system encoded by the second plasmid (pPAA3) found in these strains. This may enable the bacteria to gain access to specific sites in the host thus allowing for a more disseminated infection to take place. Indeed, we may contemplate that the bacteraemic dissemination of *P. asymbiotica* around the bodies of patients with Photorhabdosis is facilitated by the invasion of macrophage, neutrophils or other circulating white blood cells. Invasion into host cells can not only provide a niche sheltered from certain aspects of immunity, but can also grant access to certain nutrients, promoting persistence in the host.

Similarly to insect cells, following infection with *P. asymbiotica*, mammalian macrophage-like cell lines, as well as primary human macrophages later show signs of apoptosis. The effect is more pronounced with the Australian isolates as infection with the American isolates results in apoptosis being detected much later and at a lower rate. This appears to be specific to macrophage-like cells, as after infection of HeLa no signs of programmed cell death were observed (Costa et al. 2009). Apoptosis induction occurs after addition of bacteria to cells but does not require uptake of the pathogen. Additionally, it can be triggered by the addition of bacterial culture supernatant in the absence of bacteria. This indicates that the virulence factor responsible for host cell apoptosis is secreted. The identity of this virulence factor remains elusive and it is possible that there are multiple molecules exerting these effects, but one likely candidate is Mcf1. Even though Mcf1 was previously shown to cause apoptosis in HeLa cells (Dowling et al. 2004), which was not observed with *P. asymbiotica*, *mcf1* becomes up-regulated when the bacteria are grown at 37 °C, compared to expression at the normal growth temperature of 28 °C (Mulley et al. 2015). More clues come from the findings of a recent study performed with *E. coli* clones carrying cosmids that contain different regions of the *P. asymbiotica* genome (Dowling and Hodgson 2014). The effect of administration of *E. coli* lysates on macrophages was closely monitored. Here, a synergistic effect was observed between Mcf1 and the product of a nonribosomal peptide synthase (NRPS) cluster (PAU03356-PAU03357) located upstream of *mcf1*. Whilst administration of Mcf1 alone resulted in condensation of the cytoskeleton, co-administration with the NRPS product led to the formation of cytoskeletal protrusions and actin ruffles (Dowling and Hodgson 2014). This indicates that such NRPS secondary metabolites may also play a role in pathogenicity of *P. asymbiotica* as is known for several other bacterial species.

The transcription of certain other putative virulence factors of *P. asymbiotica* was shown to be up-regulated at 37 °C. These include, PVCunit4 and its associated effectors, insect toxin *pirB*, an operon containing a putative invasin (PAU_02531-37) and a *sepC* toxin-like gene (PAU_0214) (Mulley et al. 2015). On the contrary, the gene encoding the insect toxin XaxA is one of the few virulence genes that are down-regulated at 37 °C. This suggests that this factor may be specific for insect cells or would otherwise cause a problem in a human infection. Finally, the expression of the second T3SS and its effectors appears to be unaffected, however it is likely that direct contact with host cells might be a requirement for up-regulation. It should be noted that the published RNAseq work looked at changes in gene transcription in relation to temperature alone.

4.2 Metabolic Adaptations

The nutrients available in the human host niche are likely to be distinct from those in the insect host. Therefore, in addition to the deployment of toxins and other immune effector defence mechanisms, pathogens must be able to adjust their metabolism to ensure optimal nutrient acquisition and assimilation. Temperature can be sensed by many bacteria, and used as a signal to determine entry into a mammalian host. The bacteria can then respond accordingly, changing the expression of not only virulence but also metabolic genes. To measure these responses a set of transcriptomic, metabolomic and proteomic assays were performed with *P. asymbiotica* grown at the environmental temperature of 28 °C or the mammalian host temperature of 37 °C (Mulley et al. 2015). These experiments revealed that the metabolic abilities of *P. asymbiotica* undergo a large shift when grown at the elevated temperature. For example, genes encoding products involved in carbohydrate uptake and metabolism are down-regulated at 37 °C. The decrease in transcript levels of genes in these pathways is in accordance with a reduction of protein levels and the ability to metabolise these compounds in culture. Overall, *P. asymbiotica* becomes restricted in the substrates it can use as a source of carbon at 37 °C and this could reflect their availability, or lack thereof, in the new host. Examples of carbon sources it can still utilise at 37 °C are glycyl-L-proline, a product of collagen degradation and hence possibly readily available in mammals and *N*-acetyl-D-glucosamine, a component of connective tissue in both mammals and insects. On the other hand *P. asymbiotica* can use L-serine as a carbon source at both temperatures, whilst it relies on only a few amino acids and peptides to provide a source of nitrogen at 37 °C. In accordance with this, several genes encoding peptidases and amino acid and peptide transporters become up-regulated at 37 °C.

This suggests that the bacteria may rely on the degradation of host proteins and the uptake of peptides and amino acids to be used as nutrient sources in the human. Again, the secreted metalloprotease PrtA might play a role in this. The gene encoding PrtA becomes highly up-regulated at 37 °C and it has been previously suggested that the function of the protease is to support nutrient acquisition even in the insect infection since its levels rise late in the infection process (Silva et al. 2002; Daborn et al. 2001). It is tempting to speculate that a requirement for degrading host proteins at 37 °C might lead to the clinical symptoms of tissue necrosis and ulceration seen in *P. asymbiotica* infection.

Furthermore, it has been suggested that depleting certain nutrients, such as specific amino acids, from the host can have adverse effects on host cell homeostasis or immune function. This phenomenon has been coined ‘nutritional virulence’ and several examples have been seen in well-studied human pathogens (Abu Kwaik and Bumann 2013). For example, reduction in host cell serine levels would deprive host cells of a source for NADPH generation (Fan et al. 2014), whilst depletion of glutamine, which is required for immune cell function, is known to dampen immune responses (Calder and Yaqoob 1999). Glutamine is used as a source of energy by fast dividing cells such as lymphocytes and is also needed for the synthesis of other amino acids and nucleotides. Additionally, it is required for the synthesis of glutathione, which is important for protection against reactive oxygen species. Many of these pathways are connected to the conversion of glutamine into glutamate (Newsholme et al. 2003). Glutamine deprivation would therefore be detrimental to immune cells not only because it would signify a loss of energy source, but also because of lowered resistance to oxidative damage. This phenomenon has been observed in *Helicobacter pylori*, which converts both extracellular glutamine and glutathione into glutamate that it then takes up and uses mainly as an energy source (Shibayama et al. 2007). This seems to be a major contributing factor to the pathology of the bacterium (De Bruyne et al. 2016). Glutamine and glutamate are, together with tyrosine, the only amino acids *P. asymbiotica* can use as a nitrogen source at 37 °C (Mulley et al. 2015). It is therefore possible that the *P. asymbiotica* metabolism might result in glutamine depletion in host cells and have similar associated effects.

Additionally, *P. asymbiotica* is able to use purines as carbon sources and xanthine as a nitrogen source at 37 °C, whilst at the same time it down-regulates the de novo purine biosynthesis pathway. This suggests that purines are either abundant or readily available in the host and this could be particularly relevant in an intracellular setting. Indeed some pathogens like *H. pylori*, *Chlamydia trachomatis* or *Ricketsia prowazekii*, which lack either parts of or the entire purine biosynthesis pathway, need to take up nucleotides from the host (Audia and Winkler 2006; Liechti and Goldberg 2012; Tjaden et al. 1999). Additionally, xanthine depletion may also be interfering with the host immune response as xanthine is a substrate of xanthine oxidoreductase (XO) whereby it is converted into uric acid with the associated release of hydrogen peroxide (Crane et al. 2013; Martin et al. 2004). Moreover, in hypoxic conditions XO can use xanthine as a reducing agent to generate nitric oxide

(Godber et al. 2000). Thus, the ability of *Photorhabdus* to utilise xanthine as a carbon source may not only be providing the bacteria with necessary nutrients, but also acting as yet another defence strategy against the immune system.

Iron restriction is used by many animal hosts as an immune-defence strategy to limit the growth of microorganisms (Ganz and Nemeth 2015; Schaible and Kaufmann 2004). Thus, bacterial pathogens often have mechanisms to overcome this restriction (Schaible and Kaufmann 2004; Fischbach et al. 2006). In accordance with this, *P. asymbiotica* up-regulates the iron compound ABC transporter (PAU_03286), the outer membrane siderophore receptor CjrC and a bacterioferritin co-migratory protein Bcp (Mulley et al. 2015). Interestingly, the strategies employed by the human body to limit iron availability to bacteria will have different effects depending on whether the pathogens are extracellular or intracellular. So, while the sequestration of ferric ions or siderophore-bound iron will limit its availability to most pathogens, inhibition of iron release into the plasma via the degradation of ferroportin leads to a rise in its concentration inside cells like macrophages. This could lead to increased iron availability for pathogens that can survive inside these cells and so additional strategies are required to ensure that iron is excluded from intracellular sites where microbes reside, such as phagosomes (Ganz and Nemeth 2015; Schaible and Kaufmann 2004). Thus, the ability of *P. asymbiotica* to act as a facultative intracellular pathogen may be enabling it to overcome yet another human defence mechanism.

5 Conclusions

The shift of *P. asymbiotica* to a new host seems to have been facilitated by three general mechanisms. The first involves making use of virulence mechanisms already present and which are used by all *Photorhabdus* species when they infect an insect host. These could be toxins or molecules that provide defence against the host immune system. The second mechanism involves the acquisition of additional virulence factors that are appropriate for the mammalian host infection strategy. For example, specific 'toxin' effectors, which exert their actions inside host cells, like ExoU or PaTox. It can be envisaged, however, that without the aforementioned preexisting adaptations the newly acquired virulence factors would not be sufficient to confer full pathogenicity to the bacteria. Finally, *P. asymbiotica* appears to be able to adapt its metabolism to best utilise the new niche. It is possible that the changes that are observed in its metabolism upon shift to 37 °C do not exactly reflect what occurs when the bacteria are found in a human host as additional signals might be present during an infection. Even so, similar experiments with other pathogens have shown a good correlation between such in vitro data and transcriptomic data from infected cells (Bent et al. 2015; Ren and Prescott 2003). Moreover, temperature has been shown to increase expression of known virulence genes in a variety of pathogens (Konkel and Tilly 2000), sometimes controlling translation directly by altering the structure of messenger RNA molecules (Guijarro et al. 2015). Whether the metabolic shift seen with *P. asymbiotica* 37 °C is simply a result of differential availability of nutrients in the new environment, poor adaptation of certain bacterial enzymes and pathways or rather a specific strategy to use nutritional virulence to simultaneously manipulate the host, is an interesting question to consider. Finally, the inability of *P. asymbiotica* to infect insects at 37 °C despite an up-regulation of virulence genes, suggests that the changes it undergoes at this elevated temperature constitute specific adaptations to the mammalian host. We argue this failure is most likely due to the large restriction in metabolic abilities.

An analysis of *P. asymbiotica* genome sequences, in addition to temperature dependant differential transcriptomic, proteomic and metabolic studies have shed light on the characteristics that allow the members of the species to infect humans. Nevertheless, several important questions regarding their adaption to human pathogenicity remain unanswered. One obvious question that arises from comparison of the *P. asymbiotica* genome with that of the other *Photorhabdus* species is the role of the plasmid observed in *P. asymbiotica*. The fact that the European isolates HIT and JUN are unable to grow at elevated temperatures and also seem to contain a pPAU1-like plasmid, but in much lower copy numbers, and at least in one case with an altered origin of replication, hints at a role in the transition to human pathogenicity. Nevertheless, transcriptomics has not shown any significant difference in pPAU1 plasmid gene expression levels between 28 and 37 °C (Mulley et al. 2015), though the possibility that additional signals are required for gene up-regulation cannot be overlooked. One hypothesis for its function is that the plasmid may be introducing an additional level of transcriptional control over chromosomal genes. Such crosstalk between chromosome and plasmid has been observed in some other bacterial pathogens (Chitlaru et al. 2006; Zhu et al. 2010; Letek et al. 2010). This may allow better integration of the horizontally acquired genes enabling utilisation of a new niche. Furthermore, if such regulatory rewiring has occurred, it could explain why the *P. asymbiotica* ATCC43949 pPAU1 plasmid could not be cured even in laboratory conditions.

Other important questions involve the mode of transmission to humans and the presence, if any, of a mammalian host reservoir. The working hypothesis is that the nematode vector may penetrate human skin and release the bacteria; however, this has not been demonstrated so far. Certain nematodes such *Strongyloides stercoralis* or the hookworm *Ancylostoma duodenale* are known to possess this ability (Hotez et al. 2004), but this is not thought to be the case for *Heterorhabditis*. We have, however, observed that *Heterorabditis* nematodes can penetrate rat dermis ex vivo (unpublished results). Since human body temperature does not allow *Heterorhabditis* to survive (Sunanda 2009), such an infection would result in a dead-end for the nematode. Even so, the temperature on the surface of human skin at the extremities is well below 37 °C and this could allow introduction of the bacteria by the nematode. Finally, the possibility that a mammal with a lower body temperature, such as a marsupial, or poikilothermic animals like reptiles act as the reservoir for *P. asymbiotica* cannot be excluded.

References

Abu Kwaik Y, Bumann D (2013) Microbial quest for food in vivo: "Nutritional virulence" as an emerging paradigm. Cell Microbiol 15:882–890. doi:10.1111/cmi.12138

Ahantarig A, Chantawat N, Waterfield NR et al (2009) PirAB toxin from *Photorhabdus asymbiotica* as a larvicide against dengue vectors. Appl Environ Microbiol 75:4627–4629

Audia JP, Winkler HH (2006) Study of the five *Rickettsia prowazekii* proteins annotated as ATP/ADP translocases (Tlc): only Tlc1 transports ATP/ADP, while Tlc4 and Tlc5 transport other ribonucleotides. J Bacteriol 188:6261–6268. doi:10.1128/JB.00371-06

Bent ZW, Poorey K, Brazel DM et al (2015) Transcriptomic analysis of *Yersinia enterocolitica* biovar 1B infecting murine macrophages reveals new mechanisms of extracellular and intracellular survival. Infect Immun 83:2672–2685. doi:10.1128/IAI.02922-14

Bleakley BH, Chen X (1999) Survival of insect pathogenic and human clinical isolates of *Photorhabdus luminescens* in previously sterile soil. Can J Microbiol 45:273–278. doi:10.1139/cjm-45-3-273

Bondi M, Messi P, Sabia C et al (1999) Antimicrobial properties and morphological characteristics of two *Photorhabdus luminescens* strains. New Microbiol 22:117–127

Bowen DJ, Ensign JC (1998) Purification and characterization of a high-molecular-weight insecticidal protein complex produced by the entomopathogenic bacterium *Photorhabdus luminescens*. Appl Environ Microbiol 64:3029–3035

Browne N, Heelan M, Kavanagh K (2013) An analysis of the structural and functional similarities of insect hemocytes and mammalian phagocytes. Virulence 4:597–603. doi:10.4161/viru.25906

Brugirard-Ricaud K, Duchaud E, Givaudan A et al (2005) Site-specific antiphagocytic function of the *Photorhabdus luminescens* type III secretion system during insect colonization. Cell Microbiol 7:363–371. doi:10.1111/j.1462-5822.2004.00466.x

Buchmann K (2014) Evolution of innate immunity: clues from invertebrates via fish to mammals. Front Immunol 5:1–8. doi:10.3389/fimmu.2014.00459

Calder PC, Yaqoob P (1999) Glutamine and the immune system. Amino Acids 17:227–241

Chitlaru T, Gat O, Gozlan Y et al (2006) Differential proteomic analysis of the *Bacillus anthracis* secretome: distinct plasmid and chromosome CO_2-dependent cross talk mechanisms modulate extracellular proteolytic activities. J Bacteriol 188:3551–3571. doi:10.1128/JB.188.10.3551-3571.2006

Costa SCP, Girard PA, Brehélin M, Zumbihl R (2009) The emerging human pathogen *Photorhabdus asymbiotica* is a facultative intracellular bacterium and induces apoptosis of macrophage-like cells. Infect Immun 77:1022–1030. doi:10.1128/IAI.01064-08

Crane JK, Naeher TM, Broome JE, Boedeker EC (2013) Role of host xanthine oxidase in infection due to enteropathogenic and shiga-toxigenic *Escherichia coli*. Infect Immun 81:1129–1139. doi:10.1128/IAI.01124-12

Crawford JM, Portmann C, Zhang X et al (2012) Small molecule perimeter defense in entomopathogenic bacteria. Proc Natl Acad Sci 109:10821–10826. doi:10.1073/pnas.1201160109

Daborn PJ, Waterfield N, Blight M a, Ffrench-Constant RH (2001) Measuring virulence factor expression by the pathogenic bacterium *Photorhabdus luminescens* in culture and during insect infection. J Bacteriol 183:5834–5839. doi:10.1128/JB.183.20.5834-5839.2001

Daborn PJ, Waterfield N, Silva CP et al (2002) A single Photorhabdus gene, makes caterpillars floppy (mcf), allows *Escherichia coli* to persist within and kill insects. Proc Natl Acad Sci U S A 99:10742–10747. doi:10.1073/pnas.102068099

De Bruyne E, Ducatelle R, Foss D et al (2016) Oral glutathione supplementation drastically reduces *Helicobacter*-induced gastric pathologies. Sci Rep 6:20169. doi:10.1038/srep20169

Dowling AJ, Hodgson DJ (2014) An unbiased method for clustering bacterial effectors using host cellular phenotypes. Appl Environ Microbiol 80:1185–1196. doi:10.1128/AEM.03290-13

Dowling AJ, Daborn PJ, Waterfield NR et al (2004) The insecticidal toxin Makes caterpillars floppy (Mcf) promotes apoptosis in mammalian cells. Cell Microbiol 6:345–353. doi:10.1046/j.1462-5822.2003.00357.x

Duchaud E, Rusniok C, Frangeul L et al (2003) The genome sequence of the entomopathogenic bacterium *Photorhabdus luminescens*. Nat Biotechnol 21:1307–1313. doi:10.1038/nbt886

Eleftherianos I, Millichap PJ, ffrench-Constant RH, Reynolds SE (2006) RNAi suppression of recognition protein mediated immune responses in the tobacco hornworm *Manduca sexta* causes increased susceptibility to the insect pathogen *Photorhabdus*. Dev Comp Immunol 30:1099–1107. doi:10.1016/j.dci.2006.02.008

Fabbri A, Travaglione S, Fiorentini C (2010) *Escherichia coli* cytotoxic necrotizing factor 1 (CNF1): Toxin biology, in vivo applications and therapeutic potential. Toxins (Basel) 2:283–296. doi:10.3390/toxins2020282

Fan J, Ye J, Kamphorst JJ et al (2014) Quantitative flux analysis reveals folate-dependent NADPH production. Nature 510:298–302. doi:10.1038/nature13236

Farmer JJ, Jorgensen JH, Grimont PA et al (1989) *Xenorhabdus luminescens* (DNA hybridization group 5) from human clinical specimens. J Clin Microbiol 27:1594–1600

Felföldi G, Marokházi J, Képiró M, Venekei I (2009) Identification of natural target proteins indicates functions of a serralysin-type metalloprotease, PrtA, in anti-immune mechanisms. Appl Environ Microbiol 75:3120–3126. doi:10.1128/AEM.02271-08

Fischbach M a, Lin H, Liu DR, Walsh CT (2006) How pathogenic bacteria evade mammalian sabotage in the battle for iron. Nat Chem Biol 2:132–138. doi:10.1038/nchembio771

Fischer-Le Saux M, Viallard V, Brunel B et al (1999) Polyphasic classification of the genus *Photorhabdus* and proposal of new taxa: *P. luminescens* subsp. *luminescens* subsp. nov., *P. luminescens* subsp. *akhurstii* subsp. nov., *P. luminescens* subsp. *laumondii* subsp. nov., *P. temperata* sp. nov., *P.temperata* subsp. *temperata* subsp. nov. and *P. asymbiotica* sp. nov. Int J Syst Bacteriol 49(Pt 4):1645–1656. doi:10.1099/00207713-49-4-1645

Ganz T, Nemeth E (2015) Iron homeostasis in host defence and inflammation. Nat Rev Immunol 15:500–510. doi:10.1038/nri3863

Gatsogiannis C, Lang AE, Meusch D et al (2013) A syringe-like injection mechanism in *Photorhabdus luminescens* toxins. Nature 495:520–523. doi:10.1038/nature11987

Gerrard JG, Vohra R, Nimmo GR (2003) Identification of *Photorhabdus asymbiotica* in cases of human infection. Commun Dis Intell Q Rep 27:540–541

Gerrard J, Waterfield N, Vohra R, ffrench-Constant R (2004) Human infection with *Photorhabdus asymbiotica*: an emerging bacterial pathogen. Microbes Infect 6:229–237. doi:10.1016/j.micinf.2003.10.018

Gerrard JG, Joyce SA, Clarke DJ et al (2006) Nematode symbiont for *Photorhabdus asymbiotica*. Emerg Infect Dis 12:1562–1564. doi:10.3201/eid1210.060464

Gerrard JG, Waterfield NR, Sanchez-Contreeras M (2011) *Photorhabdus asymbiotica*: shedding light on a human pathogenic bioluminescent bacterium. Clin Microbiol Newsl 33:103–109. doi:10.1016/j.clinmicnews.2011.06.004

Godber BLJ, Doel JJ, Sapkota GP et al (2000) Reduction of nitrite to nitric oxide catalyzed by xanthine oxidoreductase. J Biol Chem 275:7757–7763. doi:10.1074/jbc.275.11.7757

Guijarro JA, Cascales D, García-Torrico AI et al (2015) Temperature-dependent expression of virulence genes in fish-pathogenic bacteria. Front Microbiol 6:1–11. doi:10.3389/fmicb.2015.00700

Hotez PJ, Brooker S, Bethony JM et al (2004) Hookworm Infection. N Engl J Med 351:799–807. doi:10.1056/NEJMra032492

Jank T, Bogdanović X, Wirth C et al (2013) A bacterial toxin catalyzing tyrosine glycosylation of Rho and deamidation of Gq and Gi proteins. Nat Struct Mol Biol 20:1273–1280. doi:10.1038/nsmb.2688

Joyce SA, Brachmann AO, Glazer I et al (2008) Bacterial biosynthesis of a multipotent stilbene. Angew Chemie—Int Ed 47:1942–1945. doi:10.1002/anie.200705148

Kimbrell DA, Beutler B (2001) The evolution and genetics of innate immunity. Nat Rev Genet 2:256–267. doi:10.1038/35066006

Kondakova AN, Kirsheva NA, Shashkov AS et al (2011) Low structural diversity of the *O*-polysaccharides of *Photorhabdus asymbiotica* subspp. *asymbiotica* and *australis* and their similarity to the *O*-polysaccharides of taxonomically remote bacteria including *Francisella tularensis*. Carbohydr Res 346:1951–1955. doi:10.1016/j.carres.2011.05.019

Konkel ME, Tilly K (2000) Temperature-regulated expression of bacterial virulence genes. Microbes Infect 2:157–166. doi:10.1016/S1286-4579(00)00272-0

Kuwata R, Yoshiga T, Yoshida M, Kondo E (2008) Mutualistic association of *Photorhabdus asymbiotica* with Japanese heterorhabditid entomopathogenic nematodes. Microbes Infect 10:734–741. doi:10.1016/j.micinf.2008.03.010

Lee C-T, Chen I-T, Yang Y-T et al (2015) The opportunistic marine pathogen *Vibrio parahaemolyticus* becomes virulent by acquiring a plasmid that expresses a deadly toxin. Proc Natl Acad Sci 112:10798–10803. doi:10.1073/pnas.1503129112

Letek M, González P, Macarthur I et al (2010) The genome of a pathogenic rhodococcus: cooptive virulence underpinned by key gene acquisitions. PLoS Genet. doi:10.1371/journal.pgen.1001145

Li J, Chen G, Wu H, Webster JM (1995) Identification of two pigments and a hydroxystilbene antibiotic from *Photorhabdus luminescens*. Appl Environ Microbiol 61:4329–4333

Liechti G, Goldberg JB (2012) *Helicobacter pylori* relies primarily on the purine salvage pathway for purine nucleotide biosynthesis. J Bacteriol 194:839–854. doi:10.1128/JB.05757-11

Mallo GV, Espina M, Smith AC et al (2008) SopB promotes phosphatidylinositol 3-phosphate formation on *Salmonella* vacuoles by recruiting Rab5 and Vps34. J Cell Biol 182:741–752. doi:10.1083/jcb.200804131

Martin HM, Hancock JT, Salisbury V, Harrison R (2004) Role of xanthine oxidoreductase as an antimicrobial agent. Infect Immun 72:4933–4939. doi:10.1128/IAI.72.9.4933-4939.2004

McLaughlin LM, Govoni GR, Gerke C et al (2009) The *Salmonella* SPI2 effector SseI mediates long-term systemic infection by modulating host cell migration. PLoS Pathog 5:e1000671. doi:10.1371/journal.ppat.1000671

Mulley G, Beeton ML, Wilkinson P et al (2015) From insect to man: *Photorhabdus* sheds light on the emergence of human pathogenicity. PLoS ONE 10:e0144937. doi:10.1371/journal.pone.0144937

Newsholme P, Procopio J, Ramos Lima MM et al (2003) Glutamine and glutamate—their central role in cell metabolism and function. Cell Biochem Funct 21:1–9. doi:10.1002/cbf.1003

Pankhaniya RR, Tamura M, Allmond LR et al (2004) *Pseudomonas aeruginosa* causes acute lung injury via the catalytic activity of the patatin-like phospholipase domain of ExoU. Crit Care Med 32:2293–2299

Park K-S, Ono T, Rokuda M et al (2004) Functional characterization of two type III secretion systems of *Vibrio parahaemolyticus*. Infect Immun 72:6659–6665. doi:10.1128/IAI.72.11.6659-6665.2004

Peat SM, Ffrench-Constant RH, Waterfield NR et al (2010) A robust phylogenetic framework for the bacterial genus *Photorhabdus* and its use in studying the evolution and maintenance of bioluminescence: a case for 16S, gyrB, and glnA. Mol Phylogenet Evol 57:728–740. doi:10.1016/j.ympev.2010.08.012

Plichta KL, Joyce SA, Clarke D et al (2009) *Heterorhabditis gerrardi* n. sp. (Nematoda: Heterorhabditidae): the hidden host of *Photorhabdus asymbiotica* (Enterobacteriaceae: gamma-Proteobacteria). J Helminthol 83:309–320. doi:10.1017/S0022149X09222942

Ren J, Prescott JF (2003) Analysis of virulence plasmid gene expression of intra-macrophage and in vitro grown *Rhodococcus equi* ATCC 33701. Vet Microbiol 94:167–182. doi:10.1016/S0378-1135(03)00099-3

Sato H, Frank DW, Hillard CJ et al (2003) The mechanism of action of the *Pseudomonas aeruginosa*-encoded type III cytotoxin, ExoU. EMBO J 22:2959–2969. doi:10.1093/emboj/cdg290

Schaible UE, Kaufmann SHE (2004) Iron and microbial infection. Nat Rev Microbiol 2:946–953. doi:10.1038/nrmicro1046

Shao F, Merritt PM, Bao Z et al (2002) A *Yersinia* effector and *a Pseudomonas* avirulence protein define a family of cysteine proteases functioning in bacterial pathogenesis. Cell 109:575–588. doi:10.1016/S0092-8674(02)00766-3

Shibayama K, Wachino JI, Arakawa Y et al (2007) Metabolism of glutamine and glutathione via γ-glutamyltranspeptidase and glutamate transport in *Helicobacter pylori*: possible significance in the pathophysiology of the organism. Mol Microbiol 64:396–406. doi:10.1111/j.1365-2958.2007.05661.x

Shikuma NJ, Pilhofer M, Weiss GL et al (2014) Marine tubeworm metamorphosis induced by arrays of bacterial phage tail-like structures. Science 343:529–533. doi:10.1126/science.1246794

Silva CP, Waterfield NR, Daborn PJ et al (2002) Bacterial infection of a model insect: *Photorhabdus luminescens* and *Manduca sexta*. Cell Microbiol 4:329–339. doi:10.1046/j.1462-5822.2002.00194.x

Stabb EV, Reich KA, Ruby EG (2001) *Vibrio fischeri* genes hvnA and hvnB encode secreted NAD+ -glycohydrolases. J Bacteriol 183:309–317. doi:10.1128/JB.183.1.309-317.2001

Sunanda BS (2009) Effect of temperature on the life cycle of entomopathogenic nematodes. J Biol Control, Steinernema Abbasi and Heterorhabditis Indica. doi:10.18641/jbc/23/2/46013

Thanwisai A, Tandhavanant S, Saiprom N et al (2012) Diversity of *Xenorhabdus* and *Photorhabdus* spp. and their symbiotic entomopathogenic nematodes from Thailand. PLoS One. doi:10.1371/journal.pone.0043835

Tjaden J, Winkler HH, Schwöppe C et al (1999) Two nucleotide transport proteins in *Chlamydia trachomatis*, one for net nucleoside triphosphate uptake and the other for transport of energy. J Bacteriol 181:1196–1202

Vlisidou I, Eleftherianos I, Dorus S et al (2010) The KdpD/KdpE two-component system of *Photorhabdus asymbiotica* promotes bacterial survival within *M. sexta* hemocytes. J Invertebr Pathol 105:352–362. doi:10.1016/j.jip.2010.09.020

Waterfield NR, Daborn PJ, Dowling AJ et al (2003) The insecticidal toxin Makes caterpillars floppy 2 (Mcf2) shows similarity to HrmA, an avirulence protein from a plant pathogen. FEMS Microbiol Lett 229:265–270. doi:10.1016/S0378-1097(03)00846-2

Waterfield N, Kamita SG, Hammock BD, Ffrench-Constant R (2005) The *Photorhabdus* Pir toxins are similar to a developmentally regulated insect protein but show no juvenile hormone esterase activity. FEMS Microbiol Lett 245:47–52. doi:10.1016/j.femsle.2005.02.018

Waterfield NR, Sanchez-Contreras M, Eleftherianos I et al (2008) Rapid virulence annotation (RVA): identification of virulence factors using a bacterial genome library and multiple invertebrate hosts. Proc Natl Acad Sci U S A 105:15967–15972. doi:10.1073/pnas.0711114105

Weissfeld AS, Halliday RJ, Simmons DE et al (2005) *Photorhabdus asymbiotica*, a pathogen emerging on two continents that proves that there is no substitute for a well-trained clinical microbiologist. J Clin Microbiol 43:4152–4155. doi:10.1128/JCM.43.8.4152-4155.2005

Wilkinson P, Waterfield NR, Crossman L et al (2009) Comparative genomics of the emerging human pathogen *Photorhabdus asymbiotica* with the insect pathogen *Photorhabdus luminescens*. BMC Genom 10:302. doi:10.1186/1471-2164-10-302

Wilkinson P, Paszkiewicz K, Moorhouse A et al (2010) New plasmids and putative virulence factors from the draft genome of an Australian clinical isolate of *Photorhabdus asymbiotica*. FEMS Microbiol Lett 309:136–143. doi:10.1111/j.1574-6968.2010.02030.x

Yang G, Dowling AJ, Gerike U et al (2006) *Photorhabdus* virulence cassettes confer injectable insecticidal activity against the wax moth. J Bacteriol 188:2254–2261. doi:10.1128/JB.188.6.2254-2261.2006

Zang Y-N, Jiang D-L, Cai L et al (2016) Use of a dose-response model to guide future clinical trial of benvitimod cream to treat mild and moderate psoriasis. Int J Clin Pharmacol Ther 54:87–95. doi:10.5414/CP202486

Zhu L, Liu X, Zheng X et al (2010) Global analysis of a plasmid-cured *Shigella flexneri* strain: new insights into the interaction between the chromosome and a virulence plasmid. J Proteome Res 9:843–854. doi:10.1021/pr9007514

A Review of Clinical Cases of Infection with *Photorhabdus Asymbiotica*

John G. Gerrard and Robert P. Stevens

Abstract The three recognised *Photorhabdus* species are bioluminescent Gram-negative bacilli of the family *Enterobacteriaceae*. They are all pathogenic to insects and form a symbiotic relationship with nematodes of the genus *Heterorhabditis. P. luminescens* and *P. temperata* are both harmless to humans whilst *P. asymbiotica*, on the other hand, is a human pathogen that is a symbiont of the newly described nematode vector, *Heterorhabditis gerrardi*. In this chapter, we review the epidemiological and clinical features of eighteen human cases of *P. asymbiotica* infection including fifteen from the published literature and three previously unreported cases. Human infection has been reported in the USA and Australia and probably occurs in other parts of Asia where it remains undocumented. Infection occurs most commonly in warmer months particularly after rainfall. Patients may have a history of recent exposure to sand or sandy soil. *P. asymbiotica* causes both locally invasive soft tissue infection and disseminated disease with bacteraemia. Soft tissue infection may be multifocal with involvement of more than one limb and the trunk. The organism is sensitive to a number of antibiotics in vitro, but treatment failures have been associated with the use of beta-lactams and aminoglycosides. We suggest treatment with a four-week course of an oral fluoroquinolone such as ciprofloxacin. The organism grows readily on standard media from specimens such as wound swabs, pus, blood and even sputum and can be identified in a clinical microbiology laboratory but the diagnosis needs to be considered. The correct diagnosis is most likely to be made where there is close cooperation between clinician and microbiologist.

J.G. Gerrard (✉)
Gold Coast University Hospital, Southport, QLD 4217, Australia
e-mail: John.Gerrard@health.qld.gov.au

R.P. Stevens
South Eastern Area Laboratory Services, Sutherland Hospital, Caringbah, NSW 2229, Australia

Current Topics in Microbiology and Immunology (2017) 402:179–192
DOI 10.1007/82_2016_56

Published Online: 15 March 2017

Contents

1 Background

The bacterium now known as *Photorhabdus asymbiotica* was identified as a human pathogen in 1979 by Farmer et al. (1989) and Fischer-Le Saux et al. (1999). In 2006, Gerrard et al. isolated a nematode that was found to harbour *P. asymbiotica.* This nematode was isolated from sandy soil at the home of an infected patient in Australia where the patient had been digging with his bare hands (Gerrard et al. 2006). Plichta et al. (2009) found that this nematode was phenotypically and genetically distinct and belonged to a novel species, called *Heterorhabditis gerrardi*. A seven gene multilocus sequence typing (MLST) scheme was used to confirm that the *P. asymbiotica* strains found in the nematode and the patient's infected hand were identical (Gerrard et al. 2006). Given the ability of other *Heterorhabditis* species to penetrate insect cuticle, it is plausible that this nematode could penetrate human skin. The *Heterorhabditis* spp. belong to the same order (*Rhabdidita*) as *Strongyloides stercoralis* which penetrates human skin as part of its natural life cycle. Other work suggests that *H. gerrardi* is indeed able to burrow through ex vitro pig dermal layers (Gerrard et al. 2011).

These findings strongly suggest that *H. gerrardi* vectors human infection with *P. asymbiotica*. However, it is not clear if *H. gerrardi* has a natural vertebrate host or whether humans represent an "innocent bystander" in a life cycle that normally involves insects. *Heterorhabditis* nematodes have not to date been identified in human clinical specimens, although no documented attempt has ever been made. The identification of symbiont *P. asymbiotica* infection is typically delayed by many days by which time wounds have usually been debrided making the isolation of any vector nematodes unlikely. However, this would be a worthwhile exercise in the future should the opportunity arise. The specific epithet *asymbiotica*, suggesting that the bacterium is "not a symbiont" is thus a historical error, having been designated before the symbiont nematode *H. gerrardi* was actually discovered.

2 Epidemiology

2.1 Age and Gender

Reports of fifteen cases of human infection with *P. asymbiotica* appear in the current literature. Here, we report an additional three cases in this chapter. The amount of available detail varies between cases. However, a summary of the eighteen cases appears in Table 1. Hereafter, these are referred to as "Case studies 1–18". Case studies 1–15 have been previously reported in the literature, whereas Case studies 16–18 appear here for the first time (see captions in Figs. 3, 4 and 5 for specific case details). Gender of the patient was documented for 17 of the 18 cases. The majority (11 of the 17) were male. Patient age was documented for 16 of the 18 cases and ranged from 11 to 90 years (mean = 53.9 ± 5.9 years).

2.2 Pre-morbid Status

Details of concurrent illness were documented in fifteen of the eighteen cases. The majority (10 of these 15) were in previously good health. Three patients had diabetes mellitus, and one patient was taking prednisone for polymyalgia rheumatica, and another patient was taking warfarin for cardiac valve replacements and atrial fibrillation.

2.3 Geographical Distribution

All eighteen documented case reports of *P. asymbiotica* infection have been from either the USA or Australia. In the USA, most (five) reports have been from the state of Texas with one case each from Pennsylvania and Maryland. In Australia, there have been eleven case reports along the eastern seaboard ranging from tropical Queensland to temperate Victoria. It is highly likely that the pathogen also exists in Asia, but there are no well-documented case reports from here. There has been a report of nematodes carrying *P. asymbiotica* strains in soil isolated from Thailand (Thanwisai et al. 2012), and at least one case of possible human infection said to have been acquired in Nepal but without any documentation in the literature. To our knowledge, there have been no reports to date of human infection with *Photorhabdus* spp. from Europe.

Table 1 Published cases of human infection with *P. asymbiotica*

Case	Year	Country	Location	Age/sex	Clinical	Source of isolate	Intercurrent illness	Occupation	Reference
1	1977	USA	Maryland	80/F	Endocarditis	Blood, skin			(Farmer et al. 1989)
2	1979	USA	Pennsylvania	72/F		Blood			(Farmer et al. 1989)
3	1984	USA	San Antonio, Tx	36/F	Disseminated bacterial infection	Submandible, abdomen	Diabetes mellitus		(Farmer et al. 1989)
4	1986	USA	San Antonio, Tx	78/M	Multifocal soft tissue infection left lower limb	Lower limb abscess and ulcer	Diabetes mellitus		(Farmer et al. 1989)
5	1987	USA	San Antonio, Tx	45/M	Multifocal soft tissue infection left lower limb	Lower limb abscess	Nil		(Farmer et al. 1989)
6	1989	USA	San Antonio, Tx		Groin infection	Groin			(Farmer et al. 1989)
7	1994	Australia	Melbourne, Vic.	11/F	Multifocal soft tissue infection (lower limbs and chest)	Pus and soft tissue biopsies	Nil		(Peel et al. 1999)
8	1998	Australia	Melbourne, Vic.	90/M	Cough and fever	Blood	Nil		(Peel et al. 1999)
9	1998	Australia	Wangaratta, Vic.	50/M	Multifocal soft tissue infection (upper and lower limbs)	Soft tissue abscesses	Nil	Golf course green keeper	(Peel et al. 1999)
10	1998	Australia	Murwillumbah, NSW	55/M	Multifocal soft tissue infections (upper and lower limbs, abdomen), pneumonia	Blood, sputum, pus, tissue	Nil	Farmer	(Peel et al. 1999)

(continued)

Table 1 (continued)

Case	Year	Country	Location	Age/sex	Clinical	Source of isolate	Intercurrent illness	Occupation	Reference
11	1999	Australia	Gold Coast, Qld	78/M	Soft tissue infection right foot and leg. Recent gardening bare foot in sandy soil.	Pus and tissue from right foot	Polymyalgia rheumatica on prednisone		(Gerrard et al. 2003)
12	2001	Australia	Gladstone, Qld	39/M	Soft tissue infection right ankle	Pus from ankle ulcer	Nil	Pest controller	(Gerrard et al. 2003)
13	2003	Australia	Beaudesert, Qld	29/F	Soft tissue infection right foot	Pus from foot	Nil	Farmer	(Gerrard et al. 2004)
14	2003	USA	Katy (Houston), Tx	54/M	Multifocal soft tissue infection chest and limbs	Blood, left breast abscess	Diabetes mellitus	Farmer	(Weissfeld et al. 2005)
15	2006	Australia	Kingscliff, NSW	49/M	Invasive soft tissue infection right hand and abscess left thigh	Right hand	Nil		(Gerrard et al. 2006)
16	2008	Australia	Sunshine Coast, Qld	15 M	Persistent soft tissue infection of left forearm	Left forearm	Nil	Student	See appendix
17	2009	Australia	Monto, Qld	M	Persistent soft tissue infection of right forearm	Right forearm	Nil	Gardener	See appendix
18	2009	Australia	Sydney, NSW	81/F	Soft tissue infections left arm, left thigh and right lower leg with relapse	Wound swab	Heart valve replacement on warfarin		See appendix

2.4 Environmental Exposure

In the six patients where an occupation was identified, all involved outdoor activity with likely exposure to soil: three were farmers, one a gardener (case study 17 detailed in Fig. 4), one a golf course green keeper and one a pest controller. The month of onset was reported in thirteen cases. Twelve began during the temperate summer or the tropical wet season (Gerrard et al. 2004). The Australian clinical cases have principally been identified in coastal regions. The vector of *P. asymbiotica*, *Heterorhabditis gerrardi* was also isolated from sandy soil behind the beach at Kingscliff, NSW (Gerrard et al. 2006), from the area in which the patient in case study 15 (Fig. 2) had been digging by hand, and we note that sandy coastal soil is the typical habitat of *Heterorhabditis* species as a whole (Hominick 2002). Interestingly, an association with exposure to sandy coastal soil has not been identified with the American cases.

2.5 Incidence of Human Infection

The true incidence of human infection with *P. asymbiotica* is unknown, probably because most cases are either missed or miss diagnosed (see Sect. 3). Sporadic cases have been identified in Australia but may not warrant publication unless some new aspect of the infection is observed. In Australia, therefore, human infection is probably under reported. In the USA, there has been only one case report since the original case series was published by Farmer et al. in (1989). It is likely that cases are not being diagnosed because of a failure by clinical microbiology laboratories to identify the pathogen correctly. Clinical laboratories now rely heavily on automated systems to identify bacterial isolates. The pathogen is not included in the databases of systems in common use such as Beckman Coulter MicroScan WalkAway, Biomerieux VITEK 2 or VITEK MS. So a clinical microbiology laboratory is likely to misidentify *P. asymbiotica* as a "coliform" unless the bench scientist is highly vigilant and/or experienced in identifying the organism (Weissfeld et al. 2005).

3 Recognition of *P. asymbiotica* in the Clinical Diagnostic Laboratory

The key features for *P. asymbiotica* identification include the presence of yellow pigment in the bacterial colonies on plates, nitrate not reduced to nitrite (unlike most other *Enterobacteriaceae*), annular haemolysis on blood plates (more pronounced when grown at room temperature), swarming on agar plates and weak bioluminescence of colonies on plates. Annular haemolysis is unusual and characteristic, with no haemolysis observed immediately around the colony then a very thin line observed about 6–12 mm from the colony edge. This is more apparent when the organism is cultured at room temperature. Bioluminescence can be detected with the naked eye if cultures are examined in a completely darkened room with 10 min of observer acclimatisation. The reader should, however, refer to other publications for a more detailed description of the laboratory characteristics of this organism (Peel et al. 1999).

4 Clinical Features

Human infection with *P. asymbiotica*, as reported in the literature, has been characterised by both localised soft tissue and disseminated infection (Table 1). The degree of clinical detail available for individual cases varies greatly between reports. Some form of soft tissue infection was described in the majority (16 of the 18 cases). Clinical descriptors used included: "erythematous lump", "erythematous nodule", "painful subcutaneous nodule", "diffuse, erythematous indurated subcutaneous swelling", "closed abscess", "painful necrotic ulcer", "fluctuant lymphadenopathy", "carbuncles" and "multiple pustular nodular lesions" (Figs. 1, 2, 3, 4 and 5).

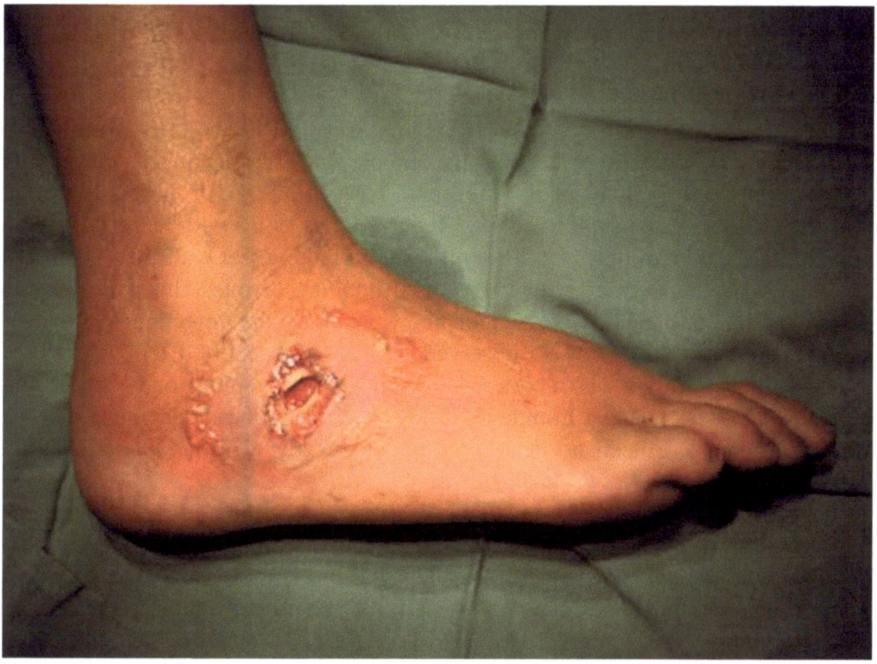

Fig. 1 Case study number 13. Twenty-nine year-old female patient with *P. asymbiotica* infection on her foot. This patient was from the town of Beaudesert in Queensland, Australia

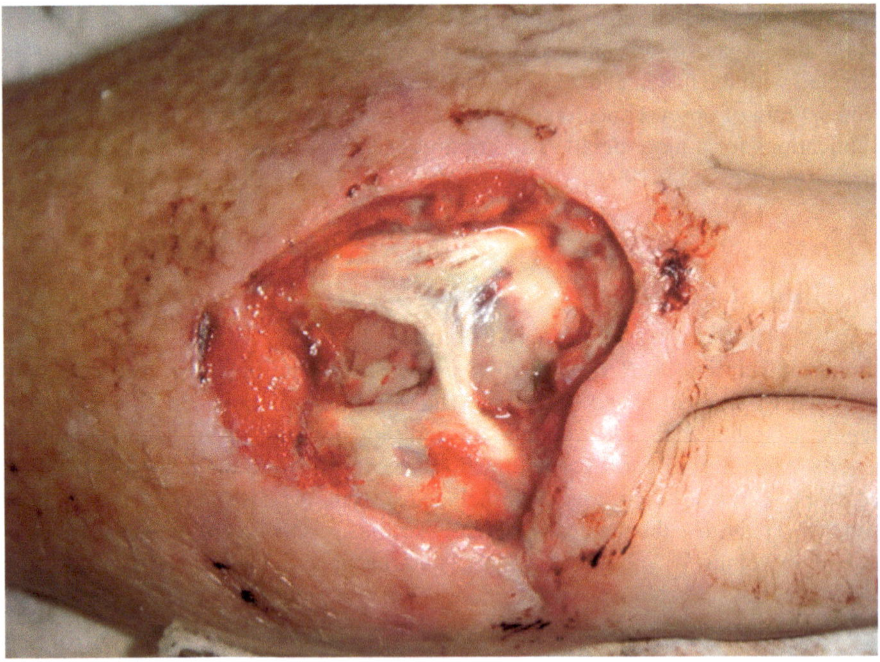

Fig. 2 Case study number 15. Forty-nine year-old male patient with an infected hand from Kingscliff, Australia. The patient had been digging a fence-post hole in sandy soil by hand. Photograph provided by Dr. David Chin (Gerrard et al. 2006)

Soft tissue infections were found to be multifocal in the majority of infections (11 of 15 cases). This could take the form of multiple abscesses on a single limb, or abscesses on multiple limbs and/or the trunk. Histopathology of biopsied skin/soft tissue lesions has demonstrated acute dermal abscesses, some with features of septic vasculitis, for example fibrin deposition in small blood vessel walls and neutrophil accumulation within vessel walls and the surrounding dermis. This appearance is though to be consistent with bacteraemic spread of the infection. Pulmonary involvement has also been also observed. Cough was reported in two patients. A left perihilar infiltrate was reported on the chest radiograph of one of these, and *P. asymbiotica* was isolated from his sputum. Bacteraemia was documented in five patients. There have been no deaths attributed directly to *P. asymbiotica* in the literature to date.

Most patients interviewed by the authors did not describe a pre-existing wound that may have been accidentally contaminated by *P. asymbiotica*. Furthermore, *Photorhabdus* spp. do not live freely in soil but are strictly associated with their vector nematodes (Bleakley and Chen 1999). These observations suggest active

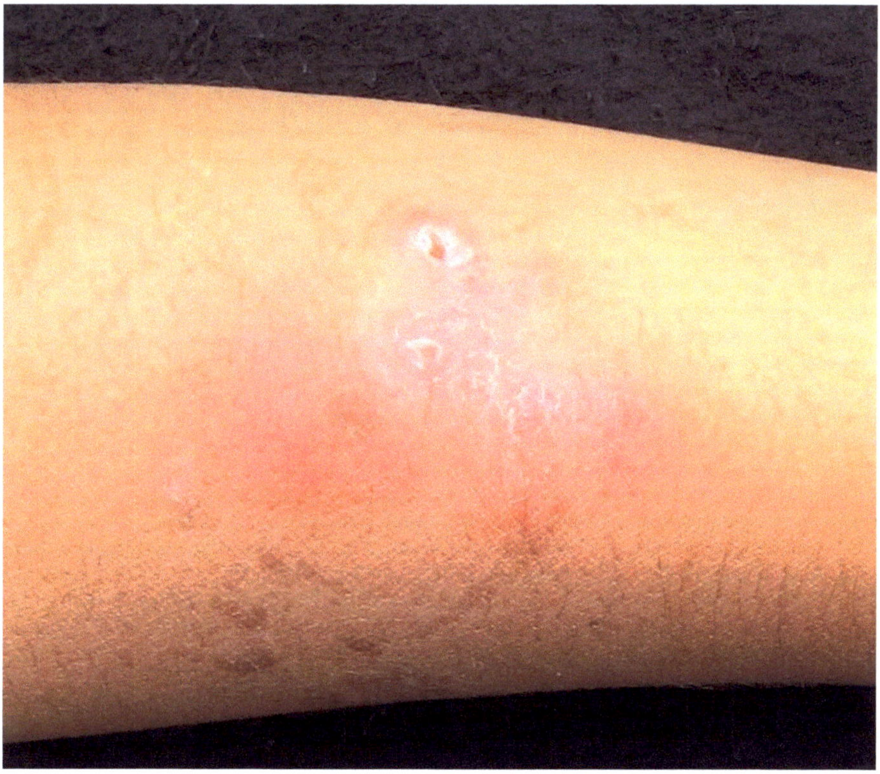

Fig. 3 Case study number 16. Fifteen year-old male patient, with a left forearm *P. asymbiotica* infection nearing resolution (*pictured*), from the Sunshine Coast of Australia. In brief, a fifteen year-old male in previously good health was diagnosed with a persistent soft tissue infection of his left forearm due to *P. asymbiotica* in 2008. He had been playing in sand near the beach in the days prior to onset. Signs and symptoms failed to resolve completely after short courses of oral amoxycillin–clavulanic acid. He was ultimately treated with a four-week course of oral ciprofloxacin that was associated with cure. This is a new case study and has not been reported elsewhere in the literature

penetration of previously intact skin by *H. gerrardi* infective juvenile (IJ) nematodes with subsequent regurgitation of *P. asymbiotica* bacteria by the IJ nematode. Notably, *Heterorhabditis* spp. belong to the same order (*Rhabdidita*) as *Strongyloides stercoralis* that penetrates human skin as part of its natural life cycle. So it seems likely that *H. gerrardi* also has the capacity to penetrate intact human skin presumably alongside its ability to penetrate its more normal insect hosts. Whether the multifocal nature of some observed soft tissue infections is the result of multiple nematode penetrations (infection by several IJ nematodes) or subsequent bacteraemic spread is still not clear.

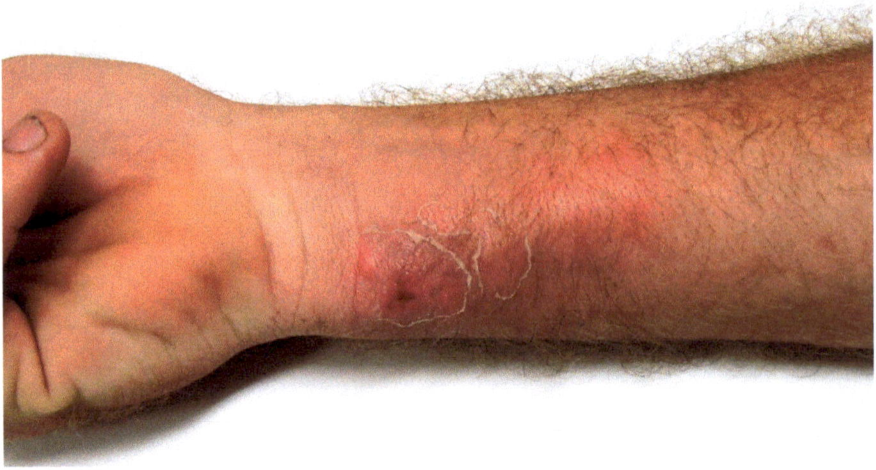

Fig. 4 Case study number 17. Male gardener with infection on wrist from Monto in Queensland, Australia. A male gardener, in previously good health, was diagnosed with a soft tissue infection of his right forearm due to *P. asymbiotica* in 2009. Note that he habitually used his right forearm to hold segments of turf grass prior to laying, suggesting that infection may have come from soil on turf. His infection resolved following a four-week course of oral ciprofloxacin. This (alongside case 16 above) is a new case study and has not been reported elsewhere in the literature

5 Treatment

P. asymbiotica isolates from the USA and Australia have been assessed for antimicrobial sensitivity using both microdilution and disk diffusion techniques. The clinical isolates were found to be sensitive in vitro to broad range of antimicrobial agents with activities against Gram-negative bacteria including quinolones, tetracyclines, aminoglycosides and β-lactams. Most were found to be resistant to ampicillin and cephalothin. In vitro sensitivity to amoxicillin–clavulanic acid seems to be affected by the temperature at which Kirby-Bauer disk diffusion is performed. At the environmental temperature (25 °C) that *P. asymbiotca* prefers, zone size may be greatly reduced. With apparent clinical failure observed using amoxycillin–clavulanic acid in some patients, an alternative regimen should probably be used.

In the absence of any form of clinical trial, it is difficult to make firm treatment recommendations. However, a relapsing course over several weeks has been documented in a number of cases. Relapse has occurred after apparently appropriate aminoglycosides or β-lactam antibiotics were administered intravenously or orally. For this reason, it has been our practice to treat with oral ciprofloxacin 500 mg b.d. for four weeks. We have given this regimen to eight (Hominick 2002) patients now with complete resolution of symptoms. The need for surgical debridement has been guided by individual clinical assessment.

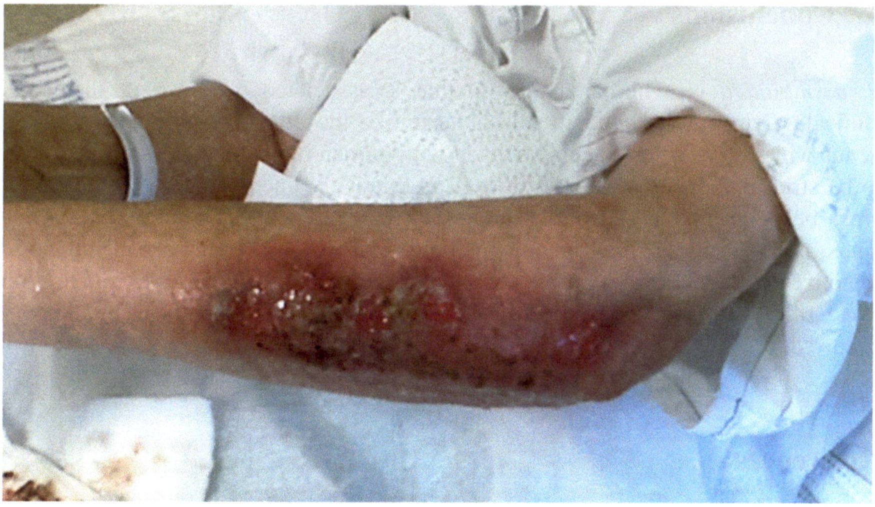

Fig. 5 Case study number 18. Eighty-one year-old female of Assyrian descent from suburban Sydney, Australia with severe infection on her left forearm. She was admitted to hospital in 2009 with soft tissue infections of her left arm, left thigh and right lower leg which were finally identified as a *P. asymbiotica* infections. She had previously undergone aortic and mitral valve replacement and was anticoagulated on warfarin. On 18 January, 2009 she presented to hospital with acute cellulitis and an abscess over the lateral epicondyle of her left elbow. This had first appeared two weeks earlier and was followed by a cutaneous abscess on the medial aspect of her left mid-thigh and another over the lower medial aspect of her right lower leg. She (like case 17 above) had a history of recent soil exposure whilst gardening with bare hands. There was no history of injury or insect bite. Her temperature on admission was 37.4 °C. She displayed a mild lymphopenia. Her erythrocyte sedimentation rate was 34 mm/hr and C-reactive protein was 38 mg/l ($N < 5$). The abscesses were drained, and she was treated with intravenous cefazolin. After 15 days, she was discharged on oral cephalexin. She represented on 10 May, 2009 with cellulitis and a recrudescence of the abscess on her elbow. The other abscesses had resolved. Note that a Gram-negative rod was originally isolated in pure growth from one of her wound swabs. This was initially misidentified as *Proteus mirabilis*. The same organism as previously isolated was recovered from swabs taken on admission. This was then confirmed to be *P. asymbiotica* using both phenotypic and molecular methods. She was then given the antibiotic ciprofloxacin which showed a rapid clinical response and cure. This (alongside cases 16 and 17 above) is a new case study and has not been reported elsewhere in the literature

Infection is cured without the need for specific therapy directed against nematodes. However, this does not in itself indicate that nematodes are inactive in the human host. For example, clinical experience with the treatment of the nematode infection onchocerciasis has demonstrated the potentially curative effect of giving antibiotics only targeting its bacterial endosymbiont *Wolbachia* (Walker et al. 2015). If clinical material could be collected early during the course of *P. asymbiotica* infection diagnosed in the future, it may prove enlightening to examine the material directly for the presence of IJ nematodes.

6 Conclusions

P. asymbiotica is a human pathogen that is associated with a very distinctive clinical syndrome. Its behaviour is quite unlike the other two species within the genus which have never been associated with human infection and therefore appear to be strict pathogens of insects only. It is relatively uncommon for pathogens to cross between animal species let alone between phyla (from arthropods to chordates). This suggests that specific pathogenic mechanisms have evolved under selective pressure. Learning more about these mechanisms and about the natural life cycle of *P. asymbiotica* and *H. gerrardi* may explain how and why this has occurred.

References

Bleakley BH, Chen X (1999) Survival of insect pathogenic and human clinical isolates of *Photorhabus luminescens* in previously sterile soil. Can J Microbiol 45:273–278

Farmer JJ, Jorgensen PA, Grimont PA, Akhurst RJ, Poinar GO, Ageron E, Pierce GV, Smith JA, Carter GP, Wilson KL et al (1989) *Xenorhabdus luminescens* (DNA Hybridization Group 5) from human clinical specimens. J Clin Microbiol 27:1594–1600

Fischer-Le Saux M, Viallard V, Brunel B, Normand P, Boemare NE (1999) Polyphasic classification of the genus *Photorhabdus* and proposal of new taxa: *P. luminescens* subsp. *luminescens* subsp. nov., *P. luminescens* subsp. *akhurstii* subsp. nov., *P. luminescens* subsp. *laumondii* subsp. nov., *P. temperata* sp. nov., *P. temperata* subsp. *tempera*. Int J Syst Bacteriol 49(Pt 4):1645–1656

Gerrard JG, McNevin S, Alfredson D, Forgan-Smith R and Fraser N (2003) *Photorhabdus* species: bioluminescent bacteria as emerging human pathogens? Emerg Infect Dis 9:251–254

Gerrard J, Waterfield N, Vohra R and ffrench-Constant R (2004) Human infection with *Photorhabdus asymbiotica*: an emerging bacterial pathogen. Microbes Infect 6:229–237

Gerrard JG, Joyce SA, Clarke DJ, ffrench-Constant RH, Nimmo GR, Looke DF, Feil EJ, Pierce L and Waterfield NR (2006) Nematode symbiont for *Photorhabdus asymbiotica*. Emerg Infect Dis 12:1562–1564

Gerrard JG, Waterfield NR and Sanchez-Contreeras M (2011) *Photorhabdus asymbiotica*: shedding light on a human pathogenic bioluminescent bacterium. Clin Microbiol Newslett 33:103–109

Hominick WM (2002) Entomopathogenic nematology. In: Gaugler R (ed) Biogeography. CAB international, London, pp 115–143

Peel MM, Alfredson DA, Gerrard JG, Davis JM, Robson JM, McDougall RJ, Scullie BL and Akhurst RJ (1999) Isolation, identification and molecular characterization of strains of *Photorhabdus luminescens* from infected humans in Australia. J Clin Microbiol 37:3647–3653

Plichta KL, Joyce SA, Clarke D, Waterfield N, Stock SP (2009) *Heterorhabditis gerrardi* n. sp. (Nematoda: Heterorhabditidae): the hidden host of *Photorhabdus asymbiotica* (Enterobacteriaceae: gamma-Proteobacteria). J Helminthol 83(4):309–320

Thanwisai A, Tandhavanant S, Saiprom N, Waterfield N, Ke Long P, Bode HB, Peacock SJ, Chantratita N (2012) Diversity of *Xenorhabdus* and *Photorhabdus* spp. and their symbiotic entomopathogenic nematodes from Thailand. PLoS One 7(9):e43835

Walker M, Specht S, Churcher TS, Hoerauf A, Taylor MJ and Basanez MG (2015) Therapeutic efficacy and microfilaricidal activity of doxycycline for the treatment of river blindness. Clin Infect Dis 60(8):1199–1207

Weissfeld AS, Halliday RJ, Simmons DE, Trevino EA, Vance PH, O'Hara CM, Sowers EG, Kern R, Koy RD, Hodde K, Bing M, Lo C, Gerrard J, Vohra R and Harper J (2005) *Photorhabdus asymbiotica*, a pathogen emerging on two continents that proves there is no substitute for a well-trained clinical microbiologist. J Clin Microbiol 43:4152–4155

Part V
Practical Guides

A Practical Guide to Recombineering in *Photorhabdus* and *Xenorhabdus*

Jia Yin, Hailong Wang, Ruijuan Li, Vinothkannan Ravichandran, Xiaoying Bian, Aiying Li, Qiang Tu, A. Francis Stewart, Jun Fu and Youming Zhang

Abstract Fluent genetic manipulation of prokaryote genomes is still limited to only a few commonly used hosts. Ideally the advanced technologies available for cloning into recombinant *Escherichia coli* should also be applicable in other prokaryotes. In particular, 'recombineering' is mediated by the lambda Red operon that permits fluent and precise engineering of the *E. coli* genome and associated recombinant DNA. The major limitation is that host-specific phage-derived

J. Yin · H. Wang · R. Li · V. Ravichandran · X. Bian · A. Li · Q. Tu · J. Fu · Y. Zhang (✉)
Helmholtz Institute of Biotechnology, State Key Laboratory of Microbial Technology, School of Life Science, Shandong University, Shanda Nanlu 27, Jinan 250100, People's Republic of China
e-mail: zhangyouming@sdu.edu.cn

J. Yin
e-mail: jiayin@sdu.edu.cn

H. Wang
e-mail: wanghailong@sdu.edu.cn

R. Li
e-mail: liruijuan@sdu.edu.cn

V. Ravichandran
e-mail: vrvinothan@sdu.edu.cn

X. Bian
e-mail: bianxiaoying@sdu.edu.cn

A. Li
e-mail: ayli@sdu.edu.cn

Q. Tu
e-mail: tuqiang1986@163.com

A. Francis Stewart · J. Fu (✉)
Department of Genomics, Dresden University of Technology, BioInnovations-Zentrum, Tatzberg 47-51, 01307 Dresden, Germany
e-mail: fujun@sdu.edu.cn

A. Francis Stewart
e-mail: francis.stewart@biotec.tu-dresden.de

Current Topics in Microbiology and Immunology (2017) 402:195–214
DOI 10.1007/82_2016_57

Published Online: 01 February 2017

recombination systems are also required in more distant species. Recently, an endogenous Red-like operon Pluγβα has been reported to be effective in both *Photorhabdus* and *Xenorhabdus* bacteria. The Pluγβα recombineering system is based on three host-specific phage proteins from *Photorhabdus luminescens*, Plu2935, Plu2936, and Plu2934, which are functional analogs of Redβ, Redα, and Redγ, respectively. In this chapter, we provide a comprehensive and up-to-date method for *P. luminescens* and *Xenorhabdus stockiae* genome engineering via the Pluγβα recombineering system. In order to facilitate the rapid construction of knock-in vectors, recET-mediated recombineering is incorporated in the pipeline. Concerted recET system in *E. coli* with Pluγβα system in *Photorhabdus* and *Xenorhabdus* could promote reverse genetics, functional genomics, and bioprospecting research for these two genera.

Contents

1 Introduction to 'Recombineering'

Photorhabdus luminescens (*P. luminescens*) is a gram-negative bacterium symbiotic with the entomopathogenic nematode (EPN) *Heterorhabditis bacteriophora*. Partial 16S rRNA gene sequences of the genera *Photorhabdus* and *Xenorhabdus* have been sequenced (Liu et al. 1997; Woese 1987), and the aligned sequences were subjected to phylogenetic analysis by maximum-likelihood and maximum-parsimony methods (Rainey and Stackebrandt 1995). Distance matrix and phylogenetic analysis suggested that these two bacterial genera are closely related despite the fact that they are associated with unrelated entomopatogenic nematodes. *P. luminescens* is highly pathogenic to insects (Lango and Clarke 2010; Glaeser and Heermann 2015), and the bacterial–EPN complex is widely used in biocontrol of insect pests. The bacteria exist in two phenotypically different forms to initiate mutualism in host nematode intestines, pathogenic variant (P form), and smaller-cell variant (M form)

(Somvanshi et al. 2012). Therefore, it is also an excellent model organism to study bacterial phenotypic heterogeneity.

P. luminescens also synthesizes an extensive array of secondary metabolites during the post-exponential phase, such as an antibiotic [3-5-dihydroxy-4-isopropylstilbene (ST)], an anthraquinone pigment (AQ), and bioluminescence (Lango and Clarke 2010; Fu et al. 2012). Moreover, there are many theoretical compounds encoded by a large number of biosynthesis gene clusters in silico analysis. However, most of these biosynthetic gene clusters are cryptic when *P. luminescens* is cultivated under laboratory conditions, such as the luminmycin (Fu et al. 2012; Bian et al. 2012b). The promoter exchange approach in *P. luminescens* resulted in indigoidine and pentadecapeptide production, which provided great insight to activate the cryptic gene cluster (Brachmanna et al. 2012; Bode et al. 2015). However, the study of these secondary metabolites is limited by the large size of the some of the gene clusters that encode them, making their cloning into recombinant *E. coli* a major stumbling block in their study (see also the chapter by Helge Bode in this volume).

An effective genome engineering tool is therefore essential to transfer and express such large blocks of genes. Two phage-encoded homologous recombination systems, which are termed 'recombinogenic engineering' or just 'recombineering,' have greatly expanded recombinant DNA technology providing accuracy and fluency in bacteria genome engineering (Datsenko and Wanner 2000; Muyrers et al. 1999; Zhang et al. 1998, 2000). They are mediated by phage-derived proteins, either by the Red operon of phage *lambda* or by recET from the Rac prophage of *E. coli*, which is a simple and efficient way to mediate homologous DNA recombination with short (about 50 b.p) homology segments (Muyrers et al. 1999; Zhang et al. 1998, 2000). The Red system, which encodes a 5′–3′ exonuclease (Redα), a single strand annealing protein (Redβ) and an exonuclease inhibitor (Redγ), can be directly applied in some gram-negative bacteria, such as *E. coli*, *Salmonella enterica* (Bunny et al. 2002), *Burkholderia* (Kang et al. 2011), and *Agrobacterium tumefaciens* (Hu et al. 2014), but host-specific phage-derived recombination systems are required in more distant species. The new recombineering systems based on host-specific phage-derived recombinases have been established in *Yersinia pseudotuberculosis* (Derbise et al. 2003), *Mycobacterium tuberculosis* (Van Kessel and Hatfull 2006, 2008), *Pseudomonas syringae* (Bao et al. 2012), *Lactococcus lactis* (van Pijkeren and Britton 2012), *Lactobacillus plantarum* (Yang et al. 2015), *Clostridium perfringens* (Dong et al. 2013), and *Bacillus subtilis* (Sun et al. 2015).

The Red system had been used to manipulate the *Photorhabdus* chromosome, but it needs a long homology arm which is beyond the convenience of oligos synthesis (Somvanshi et al. 2012; Tang et al. 2010). Pluγβα recombineering system is a more robust method for the functional genomics research in *P. luminescens* and *Xenorhabdus stockiae* (Yin et al. 2015b). The Pluγβα recombineering system is based on three host-specific phage proteins from *P. luminescens*, Plu2935, Plu2936, and Plu2934, which are functional analogs of Redβ, Redα, and Redγ, respectively. This new recombineering system therefore renders the host genome amenable to fast and accurate manipulation for reverse genetics, functional genomics, and bioprospecting.

In this chapter, we provide a comprehensive and up-to-date practical guide for *P. luminescens* and *X. stockiae* genome engineering using Pluγβα recombineering system. To facilitate the construction of knock-in vectors for Pluγβα recombineering, a method using recET to assemble multiple DNA fragments is also included. Combining recET recombineering with Pluγβα system could promote reverse genetics, functional genomics, and bioprospecting research for *P. luminescens*, *X. stockiae,* and other closely related bacteria.

2 Reagents and Equipment Needed for Practical 'Recombineering'

2.1 Reagents

- **Luria Broth (LB)** Dissolve 5 g yeast extract, 10 g tryptone and 5 g NaCl in 1 L ddH_2O, and transfer 100 or 250 mL aliquots into a glass stock bottle and autoclave. Add appropriate antibiotics before use.
- **LB agar plates** Transfer required aliquots of autoclaved LB broth into a glass stock bottle, add 1.2% agar (w/v) and dissolve it by boiling, and then pour molten medium into sterile petri dishes. Add appropriate antibiotics when the medium has cooled to 50 °C if it is required.
- **0.7% (w/v) agarose gels** Add a certain amount of agarose to appropriate volume 1 × TBE or TAE buffer (w/v = 0.7%). Completely melt the agarose in a microwave and then cool down to ~60 °C. Add 5% (v/v) ethidium bromide solution (10 mg mL^{-1}), mix thoroughly, pour the dissolved agarose into a mold with an appropriate comb and let it solidify.
- ***E. coli* strain** The *E. coli* GB2005 cells harboring pSC101-BAD-34-35-36-amp plasmid should be grown freshly on LB plates supplemented with 100 μg mL^{-1} ampicillin or LB broth supplemented with 50 μg mL^{-1} ampicillin.
- ***P. luminescens* and *X. stockiae* strain** The *P. luminescens* and *X. stockiae* cells harboring pSC101-BAD-34-35-36-amp plasmid should be grown freshly on LB plates supplemented with 100 μg mL^{-1} ampicillin or LB broth supplemented with 50 μg mL^{-1} ampicillin. After recombineering, cells should be grown freshly on LB plates supplemented with 10 μg mL^{-1} kanamycin or 3 μg mL^{-1} gentamicin.
- **Antibiotic stock solutions** Prepare stock solution of ampicillin (100 mg mL^{-1}), kanamycin (30 mg mL^{-1}), and gentamicin (50 mg mL^{-1}) in autoclaved ddH_2O. Pass the solutions through 0.2-μm syringe filters to sterilize inside a sterile laminar hood. The stock solution can be stored at −20 °C for further use. Ampicillin (100 μg mL^{-1}), gentamicin (3 μg mL^{-1}), chloramphenicol (15 μg

mL^{-1}), kanamycin (10 μg mL^{-1}), rifampicin (100 μg mL^{-1}), and apramycin (25 μg mL^{-1}) can be used in *P. luminescens* and *X. stockiae.*

- **3M sodium acetate pH 5.2** Dissolve 40.8 g sodium acetate trihydrate in ~80 mL ddH_2O and adjust to pH 5.2 with acetic acid, and finally bring to a final volume of 100 mL. The solution can be stored at room temperature (RT) for at least one year.
- **10% Glycerol** 50 mL pure glycerin is mixed with 450 mL ddH_2O. The solution can be kept at RT for several months after autoclave.
- **GH electroporation buffer** Dissolve 0.04766 g HEPES in ~80 mL sterile 10% glycerol and bring to a final volume of 100 mL. The pH should be ~6.5 without addition of acetic acid. The solution should be freshly prepared before use.
- **10% L-(+)-arabinose** Dissolve 10 g L-(+)-arabinose powder in ~80 mL ddH_2O and bring to a final volume of 100 mL. Pass the solution through 0.2-μm syringe filters to sterilize inside a sterile laminar hood. The stock solution can be stored at −20 °C for further use.
- **1 × TBE electrophoresis buffer** Dilute 10 × TBE buffer with ddH_2O to a 1 × solution and store it at RT for several months.
- **The lox71-km-lox66/-Ptet cassette** Release the cassette from pSC101-lox66-neo-lox71-Ptet-GFP by restriction digestion (*Bam*HI/*Ava*I) to be PCR template, purify with QIA quick PCR Purification Kit or ethanol precipitation, and elute in autoclaved ddH_2O. Quantify the elution by NanoDrop UV spectrophotometer. The concentration should be greater than 100 ng μL^{-1} and can be kept at −20 °C for several months. The PCR template should be diluted to 2.5 ng μL^{-1} before use.
  Note: (i) *The restriction digestion of pSC101-lox71-neo-lox66-Ptet-GFP and decreasing its concentration will avoid the selection background from carryover of the intact plasmid after* recombineering, *because the transformation efficiency a carryover plasmid might be higher than homologous recombination efficiency.* (ii) *To make unmarkered mutants of P. luminescens and X. stockiae, the Cre-lox system with the variant lox66 and lox71 sites was used to prevent the instability caused by multiple loxP sites in the genome.*
- **The lox71-genta-lox66/-Ptet cassette** The cassette can be obtained directly from the intact pR6K-Tps-lox71-Gent-lox66-Ptet-T7RP plasmid by PCR, because pR6K-Tps-lox71-Gent-lox66-PtetR-T7RP will not replicate in *P. luminescens* TT01.
- **Origin** p15A, pBR322, and pSC101 can be replicated in *P. luminescens* and *X. stockiae*
- Oligos were synthesized by Sangon Biotech (China)
- DL5000 DNA Marker (Takara, cat. -no. D503A)
- Restriction enzymes (New England Biolabs)
- PrimeSTAR Max DNA Polymerase (Takara, cat. -no. R045B)
- 10 × Tris/Boric Acid/EDTA (TBE) buffer (Bio-Rad, cat. -no. 161-070)
- Ethidium bromide solution (10 mg mL^{-1}, Dingguo, cat. -no. NEP028-1)

- Agarose (Invitrogen, cat. -no. 75510-019)
- Buffer P1 (Tiangen, cat. -no. RK101-02)
- Buffer P2 (Tiangen, cat. -no. RK102-02)
- Buffer P3 (Tiangen, cat. -no. RK103-02)
- Universal DNA Purification kit (Tiangen, cat. -no. DP214-03)
- Tryptone (OXOID, cat. -no. LP0042)
- Yeast extract (OXOID, cat. -no. LP0021)
- NaCl (Sangon Biotec, cat. -no. SB0476)
- Agar (Solarbio, cat. -no. A8190)
- Glycerol (Sigma-Aldrich, cat. -no.15523)
- HEPES (Sangon Biotec, cat. -no. A100511-0250)
- NaOH (Sangon Biotec, cat. -no. A100173-0500)
- Ampicillin sodium salt (Dingguo, cat. -no. AA022-25G)
- Kanamycin sulfate (Dingguo, cat. -no. AK177-25G)
- Gentamicin solution (50 mg mL^{-1}, Sigma-Aldrich, cat. -no. G1397)
- Apramycin sulfate salt (Dingguo, cat. -no. MB-A4588)
- Chloramphenicol (Dingguo, cat. -no. AC060-50G)
- Sodium acetate trihydrate ($NaAc{\cdot}3H_2O$; Sinopharm, cat. -no. 10018718)
- Isopropanol (Sinopharm, cat. -no. 80109218)
- Ethanol absolute (Sinopharm, cat. -no. 10009218)
- RNase A (10 mg mL^{-1}; DNase and protease free; Thermo Scientific, cat. -no. EN0531)
- L-(+)-arabinose (Sigma-Aldrich, cat. -no. A3256)

2.2 Equipment Needed for 'Recombineering'

- High-speed microcentrifuge (Sangon Biotec, cat. -no. G508009-0001)
- Mini centrifuge (Sangon Biotec, cat. -no. G508008-0001)
- Benchtop centrifuge 5424R kept at 2 °C (Eppendorf, cat. -no. 5404000.014)
- Thermo Shaker Incubator (Hangzhou Allsheng, cat. -no. MSC-100)
- Thermomixer F1.5 (Eppendorf, cat. -no. 5384000.071)
- Electroporation cuvettes with 1-mm gap kept on ice (Eppendorf, cat. -no. 940001005)
- Electroporator 2510 (Eppendorf, cat. -no. 940000009)
- Mini Vortex meter (Sangon Biotec, cat. -no. G002033-0001)
- Digital gel imaging system (Bio-Rad, GelDoc XR)
- UV spectrophotometer (Thermo Scientific, NanoDrop 2000c)
- UV–Vis spectrophotometer (T6 new century, PGeneral)
- Gel electrophoresis apparatus (Beijing Junyi)
- Multipette (Eppendorf, cat. -no. 4981000.019)
- 10-mL Combitips (Eppendorf, cat. -no. 0030089.464)
- Incubator kept at 30 and 37 °C (Ningbo Jiangnan, cat. -no. HWS-0288)

- Sterile 10-μL inoculation loop (Sangon Biotec, cat. -no. IL311-10-S-Q)
- Sterile 1-μL inoculation loop (Sangon Biotec, cat. -no. IL311-1-S)
- Petri dishes, 94 mm × 16 mm (Greiner bio-one, cat. -no. 633180)
- Syringe needle, 25G 5/8, 0.5 mm × 16 mm (BD Medical, cat. -no. 301805)
- 0.2-μm syringe filter (PALL, cat. -no. PN4612)

3 Pluγβα Recombineering in *Photorhabdus* and *Xenorhabdus*

Preliminary data indicated that the *lambda* Red proteins did not mediate efficient editing of the *P. luminescens* genome (Tang et al. 2010; Somvanshi et al. 2012), but in contrast the Pluγβα recombineering system provides precise manipulation of the host genome (Yin et al. 2015b). The impacts of specific molecular interactions of host factors on Red-mediated recombineering system therefore need to be explored further. The chaperone system GrpE–DnaK–DnaJ plays a crucial role in Red recombination and may contribute to the folding or stability of either Redβ or Redα or both (Poteete 2013). The major DNA polymerases (Pol I and Pol III) and DNA ligase are also directly involved with oligo-recombination in the host (Li et al. 2013). These results suggest that recombinases from bacterial phages, prophages, or their remnants might perform better in their native hosts.

Temperature is also an important factor for the efficient functioning of recombinases. The optimum temperature for Pluγβα is ~30 °C (Yin et al. 2015b), whereas the optimal recombination activity for Redγβα is 37 °C. Those differences in temperature optima are consistent with their host origins, as Pluγβα is derived from the *P. luminescens* bacterium while Redγβα is from the *E. coli* lambda phage.

The Cre/loxP site-specific recombination system of bacteriophage P1 is efficient for genetic manipulation in all cell types. The gene 'knock-in' is mediated by Cre as an integrative recombination (Frank and Michael 2001). The gene 'knockout' uses the excision function of the Cre when the DNA fragment is flanked by two parallel oriented loxP sites (Balbás and Gosset 2001; Sektas and Specht 2005). In this section, we present a detailed method for *P. luminescens* and *X. stockiae* genome engineering combining Pluγβα recombineering system with Cre/loxP site-specific recombination system.

3.1 Materials Needed

- P. *luminescens* TT01 (Tang et al. 2010)
- *X. stockiae* (Yin et al. 2015b)
- GB2005 (Fu et al. 2010)
- pSC101-BAD-34-35-36-amp (Yin et al. 2015b)

- pSC101-lox71-neo-lox66-Ptet-GFP (Yin et al. 2015b)
- pSC101-cl578-Cre-cm
- pR6K-Tps-lox71-Gent-lox66-Ptet-T7RP (Bian et al. 2012a)

3.2 Method

3.2.1 General Schema for the Pipeline

Figure 1 illustrates the genome engineering of *P. luminescens* and *X. stockiae*. It is based on Pluγβα recombineering in *Photorhabdus* and *Xenorhabdus*. To capacitate the *P. luminescens* and *X. stockiae* for recombineering, the recombinase proteins

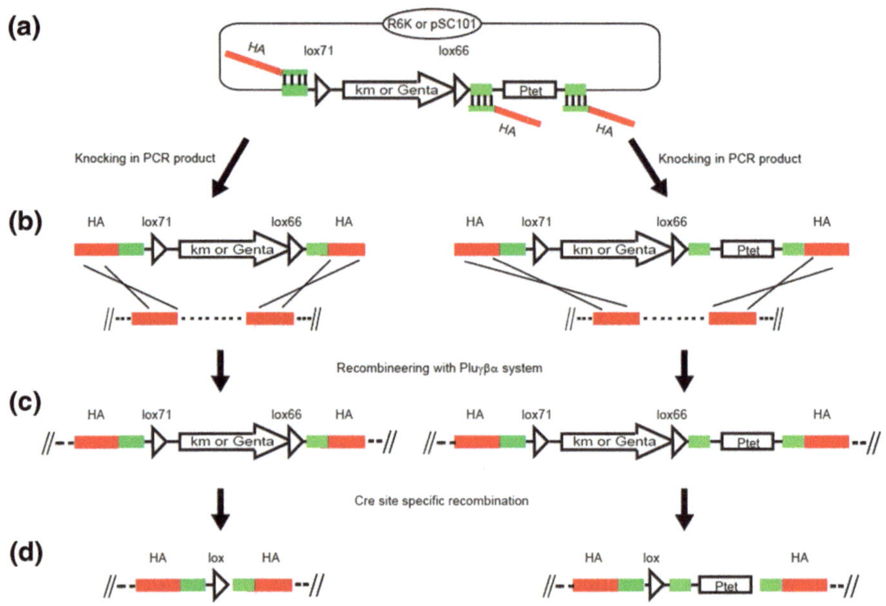

Fig. 1 Schematic diagram of *P. luminescens* and *X. stockiae* genome engineering. **a** The pSC101-lox71-neo-lox66-Ptet-GFP is used as a PCR template for the lox71-neo-lox66/-Ptet. The pR6K-Tps-lox71-Gent-lox66-Ptet-T7RP is used as a PCR template for the lox71-Gent-lox66/-Ptet. The necessary elements for knocking out or knocking in are shown in the diagram. Those plasmids serve as templates for PCR amplification using oligonucleotides that contain homology arms (HA), HA share identical terminal regions with target locus. **b** The PCR product is electroporated into *P. luminescens* or *X. stockiae* harboring the recombineering expression plasmid, pSC101-BAD-34-35-36-amp (not illustrated). The *left panel* is for knocking out PCR product, and the *right panel* is for knocking in PCR product. **c** Screen the recombinants by kanamycin or gentamicin selection. **d** Cre catalyzes excision of the DNA fragment flanked by two parallel oriented mutant lox sites (lox71-lox66) which allows for repeated genome manipulations in a single genetic background

Plu$\gamma\beta\alpha$ need to be expressed. However, it is advantageous to limit their expression to only the time interval required because this also limits the chances of unwanted recombination. Consequently, we employ the excellent properties of the arabinose inducible BAD promoter (Guzman et al. 1995), to firmly regulate Plu$\gamma\beta\alpha$ expression (Yin et al. 2015b). The pSC101 plasmid is low copy ($\sim$5 per cell) and encodes a temperature-sensitive replication protein so that it replicates at 30 °C and not at 37 °C (Hashimoto-Gotoh and Sekiguchi 1977). Hence, it can be easily eliminated from the host by temperature shift. After recombineering and colony PCR verification, the unmarkered mutant bacteria can be generated by Cre site-specific recombination.

3.2.2 Preparation of 'Knock-In' or 'Knockout' Cassette

Attaching homology arms to the cassette for knock-in or knockout is easily achieved by PCR amplification. The 95 mer oligonucleotides include 75 nucleotides (nt.) homology arms flanking the targeting region and 20 nt. PCR primers at their 3′ ends. The PCR templates are shown in Fig. 1, and the reaction conditions are in Tables 1 and 2.

The pSC101 origin plasmid can replicate in *P. luminescens*. To avoid the background from the intact plasmid, the restriction digestion of pSC101-lox71-neo-lox66-Ptet-GFP with BamHI/AvaI is required. Once digestion completed, sodium acetate and ethanol precipitation is carried out using one-tenth volume of 3M sodium acetate trihydrate and three fold volumes of ethanol. The DNA pellet is then rinsed with 70% ethanol. After air drying the DNA pellet, the digested plasmid should be resuspended in ddH_2O at $\sim$300 ng μL^{-1}. The PCR template should then be diluted to 2.5 ng μL^{-1} with ddH_2O before use.

Table 1 PCRs

Template	pSC101-lox71-neo-lox66-Ptet-GFP[a]	pR6K-Tps-lox71-Gent-lox66-Ptet-T7RP[a]
2 × PrimeSTAR Max premix	25 μL	25 μL
DMSO	1 μL	1 μL
Oligo 1 (50 pmol μL^{-1})	1 μL	1 μL
Oligo 2 (50 pmol μL^{-1})	1 μL	1 μL
Template	1 μL of 2.5 ng μL^{-1}	1 μL of 50 ng μL^{-1}
H_2O to total	50 μL	50 μL

[a]For the lox71-neo-lox66/-Ptet PCR, digest about 10 μg plasmid with BamHI/AvaI, precipitate and dissolve in ddH_2O to the final concentration of 2.5 ng μL^{-1} as PCR template. For lox71-Gent-lox66/-Ptet PCR, simply use 50 ng from standard plasmid preparations as PCR template

Table 2 Cycle setting for PCRs

lox71-neo-lox66/-Ptet or lox71-Gent-lox66/-Ptet[a]		
98 °C	2 min	
98 °C	15 s	35×
57 °C	15 s	
72 °C	30 s	
72 °C	10 min	
8 °C	End	

[a]The PCR products of lox71-neo-lox66 and lox71-Gent-lox66 are used in the genome engineering of *P. luminescens* and *X. stockiae* (knockout). The PCR products of lox71-neo-lox66-Ptet and lox71-Gent-lox66-Ptet are used for silent PKS/NRPS gene cluster expression (knock-in)

3.2.3 PCR Purification and Yield

One micro liter of the PCR products is checked by electrophoresis on 0.7% agarose gel. Rest of the yield was purified using Universal DNA Purification Kit and eluted with 30 μL ddH_2O. The purification is essential to get rid of unincorporated oligonucleotides and salt. The yield should be more than 3 μg at 200–600 ng μL^{-1}. For recombineering, 1–2 μg PCR product is needed for one electroporation.

3.2.4 Preparation of *P. luminescens* and *X. stockiae* Electrocompetent Cells

- Inoculate single colony in fresh LB with appropriate antibiotics at 30 °C for 16 h.
- The overnight cultures were diluted (starting OD_{600} around 0.15) into 1.3 mL fresh LB medium with appropriate antibiotics and grown at 30 °C, 950 rpm for 5 h.
  Note: *The optimal* OD_{600} *is around 1.1 before preparing electrocompetent cells. The cultivation time can be extended if the* OD_{600} *does not reach this value.*
  Note: *When the electrocompetent cells for recombineering is prepared, the fresh culture was grown at 30 °C, 950 rpm for 4.5 h until the* OD_{600} *was around 0.85. After the addition of L-(+)-arabinose to a final concentration of 2.5 mg* mL^{-1}*, the cells were grown at 30 °C, 950 rpm for 30 min until the* OD_{600} *was around 1.15.*
- Spin it at 9000 rpm for 30 s in cooling centrifuge at 2 °C.
- Discard the supernatant as much as possible.
- Resuspend the pellet in 1 mL of ice-cold GH buffer (10% Glycerol, 2 mM HEPES, pH = 6.5) by a vortex.
- Spin at 9200 rpm for 30 s in the cooling centrifuge at 2 °C.
- Discard the supernatant as much as possible.
- Resuspend the pellet in 1 mL of ice-cold GH buffer by a vortex.
- Spin the cells down at 9400 rpm for 30 s in the cooling centrifuge at 2 °C.

- Discard the supernatant and leave about 30 μL.
- Use the cells immediately.

3.2.5 Transformation of the Recombinase Expression Plasmid

To capacitate the *P. luminescens* and *X. stockiae* for recombineering, the plasmid pSC101-BAD-34-35-36-amp is introduced into *P. luminescens* and *X. stockiae* by electroporation.

- Add 500 ng plasmid to the electrocompetent cells and pipette the mixture into the chilled 1 mm electroporation cuvette.
- Set the electroporator to 1200 V, 10 μF, 600 Ω.
  Note: *This setting is based on an Eppendorf® Electroporator 2510 using a 1 mm gap cuvette.*
- Gently tap the cuvette on the table to remove air bubbles and dry the metallic sides of the cuvette with a tissue.
- Place the cuvette into the holder of the electroporator, insert, and push the 'pulse' button twice.
- Add 1 mL fresh LB medium to the cuvette. Mix the cells gently by pipetting up and down and transfer into the reaction tube.
- Incubate the cultures on a shaking incubator at 950 rpm for 100 min.
- Spread the cells on ampicillin (100 μg mL^{-1}) plates.

3.2.6 Identify the Transformants

Due to the strong nucleases activity in the wild-type *P. luminescens* and *X. stockiae*, the DNA yield of Mini-Prep is very low. *E. coli* GB05 (Fu et al. 2010) is an excellent host for exogenous DNA preparation. The plasmid should be isolated from *P. luminescens* and retransformed into *E. coli* GB05 for its identification.

- Inoculate the single colony in fresh LB with ampicillin (50 μg mL^{-1}) at 30 °C for 16 h.
  Note: *Pick the large colonies. The small colonies cannot grow in the liquid medium with ampicillin (50 μg mL^{-1}).*
- Spin the cells at 12,000 rpm for 1 min.
  Note: *Prepare backup tubes before starting.*
- Discard the supernatant as much as possible.
- Add 200 μL of P1 buffer (keep P1 at 4 °C because of the RNaseA in the buffer).
- Resuspend the pellet by a Thermomixer.
- Add 200 μL of P2 buffer.
  Note: Do not place P2 on ice, as it will cause precipitation.
- Mix the tubes by inversion 5 times gently.

Note: *Do not allow the lysis to proceed for more than 5 min.*

- Add 200 μL of P3 buffer.
- Mix by inversion 5 times.
- Spin the tubes at 13,200 rpm for 20 min.
  Note: *Centrifugation for two times is required for P. luminescens Mini-Prep, and the DNA will not be contaminated with protein.*
- Prepare a new 1.5-mL tube for each sample and add 500 μL of isopropanol.
- After centrifugation, take out the supernatant and transfer to isopropanol containing tube.
- Mix by shaking the tubes vigorously.
- Spin the tubes at 13,200 rpm for 20 min.
- Discard the supernatant and invert the tube on a tissue.
- Carefully add 1 mL of 70% ethanol.
  Note: *Centrifuge again for 5 min if pellet is only loosely attached to the wall of the tube.*
- Discard the ethanol and invert the tube on a tissue.
- Dry the pellet for 10–15 min on a heating block (45 °C).
  Note: *Do not over dry it; otherwise, it will be hard to dissolve.*
- Dissolve the dried DNA in 5–10 μL ddH_2O.
- Transformation of the obtained DNA into in *E. coli* GB05.
- Restriction analysis of plasmids after Mini-Prep from *E. coli* GB05.

3.2.7 Genome Engineering of the *P. luminescens* and *X. stockiae*

- Add 1 μL PCR products (lox71-neo-lox66/-Ptet or lox71-Gent-lox66/-Ptet) into *P. luminescens* and *X. stockiae* electrocompetent cells harboring the Pluγβα expression plasmid (pSC101-BAD-34-35-36-amp).
- Perform electroporation using ice-cold cuvettes and an Eppendorf 2510 electroporator set at 1200 V, 10 μF, 600 Ω.
- Add 1 μL fresh LB medium after electroporation.
- Incubate the cells for recovery at 30 °C for 100 min with shaking and then spread on kanamycin (10 μg mL^{-1}) or gentamicin (3 μg mL^{-1}) plates.
- After incubation at 30 °C for 48 h, check whether the recombinants were by colony PCR.
  Note: *For knock-in, the recombinants should be checked on both junctions by colony PCR when the insertion sequence is large.*

3.2.8 Seamless Mutagenesis of the *P. luminescens* and *X. stockiae*

In order to generate the seamless *P. luminescens* and *X. stockiae* mutant, Cre catalyzes excision of the DNA fragment flanked by two parallel oriented variant lox sites.

- Five hundred nanogram plasmid of pSC101-cl578-Cre-cm is introduced into the genome engineered bacteria by electroporation.
  Note: *cl578 is a temperature-sensitive inducible promoter, the Cre protein can be induced by a temperature shift from 30* to *37 °C.*
- Inoculate single colony in 1 mL fresh LB with chloramphenicol (10 μg mL^{-1}) at 30 °C for 6 h.
- Shift the tubes to 37 °C, 950 rpm for another 40 min.
- Plate 100 μL of the cells with a loop on LB agar plates and incubate the plates at 30 °C for 24 h.
- Pick single colony and double streak on an antibiotic selection plate (kanamycin or gentamicin) and a LB plate without antibiotic.
- Select kanamycin or gentamicin sensitive strain for colony PCR verification.
- Plasmid of pSC101-cl578-Cre-cm is removed by increasing the temperature at 37 °C.

4 Rapid Construction of Knock-in Vectors Using the recET System

The genomic integration of large sections of foreign DNA (e.g., an inducible promoter or metabolite synthesis pathway) in an organism has been reported in numerous prokaryotes (Gross et al. 2006; Yin et al. 2015a), yeast (Chandran et al. 2014), zebrafish (Kimura et al. 2014), and even mammals (Li et al. 2016). The targeting construct (a suicide vector or linear DNA fragment) is introduced into the host cells, and the exogenous DNA could be integrated into the loci of interest by homologous recombination. The targeting events are then screened by appropriate selection and confirmed by genotyping which is usually PCR or Southern Blotting. The foreign DNA with a selection marker can be obtained from a PCR, as illustrated in the last section. However, PCR has some limitations for the generation of larger fragments.

For the genomic integration of a large section of foreign DNA with multiple elements, DNA assembly technologies are required to obtain targeting constructs. RecE/RecT encoded by the Rac prophage has been used to promote cloning by homologous recombination. RecE is a 5′–3′ exonuclease that generates 3′-ended ssDNA overhangs (Carter and Radding 1971; Little 1967). RecT is a single strand annealing protein (SSAP) that binds to ssDNA and forms a nucleoprotein filament with complementary ssDNA. This strategy increases the efficiency of recombination between homologous ends $\geq$35 bp. Recombineering in *E. coli* mediated by recET was developed to clone large biosynthetic gene clusters from a complex DNA source into a vector by linear plus linear homologous recombination (LLHR) (Fu et al. 2012) and assemble multiple fragments in vivo in single reaction (Yin et al. 2015a). In this section, we will present a method to construct a target vector using recombineering for *P. luminescens* and *X. stockiae*.

4.1 Materials

- GB05-dir (Fu et al. 2012)
- p15A-cm-ccdB (Wang et al. 2014)
- pSC101-tet
- pSC101-lox71-neo-lox66-Ptet-GFP (Yin et al. 2015b)
- pR6K-Tps-lox71-Gent-lox66-Ptet-T7RP (Bian et al. 2012a)

4.2 Method

4.2.1 General Scheme of the Pipeline

A schematic diagram for constructing a knock-in vector is illustrated in Fig. 2. It is based on recombineering system in *E. coli* GB05-dir host (Fu et al. 2012). The *recE, recT,* and *redγ* genes are integrated into the chromosome and expressed from the arabinose-responsive BAD promoter. Multiple fragments can be assembled into one target construct by homologous recombination. The constructed plasmid can be suicidal by using a host-specific replication origin R6K or can be conditional suicide by using a temperature-sensitive replication origin pSC101.

4.2.2 Fragment Assembly

- Attaching homology arms to the linear cloning vector is easily achieved by PCR amplification. The 120 mer oligonucleotides include 60 nt. homology arms to the targeting locus, 40 nt. homology arms to the cassettes, and 20 nt. PCR primers at their 3′ ends.
- Cassettes can be obtained from PCR or digestion from other plasmids by restriction enzyme.

  Note: (1) *The homology arm can be attached through oligonucleotides synthesis when the cassette is obtained from PCR.* (2) *There isn't great impaction on recombineering efficiency if the homology arm is not completely exposed when the cassette is obtained from digested plasmid.*

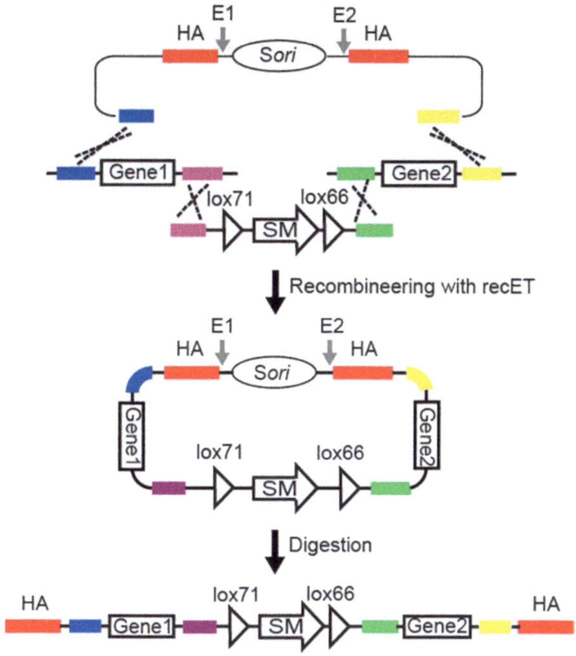

Fig. 2 Strategy diagram for constructing a knock-in vector. Multiple fragments could be assembled into a targeting construct by recET-mediated recombineering. Subsequently it can be integrated into the target locus after exposing the homology arm by restriction digestion. S*ori* is suicide origin. Cross-region is homology arm for constructing targeting vector, and HA is homology arm that share identical terminal regions with target locus. SM is selection marker, e.g., gentamicin and kanamycin resistance genes

- Attaching homology arms to the selection marker including the flanked lox71 and lox66 sites is easily achieved by PCR amplification. The template is pSC101-lox71-neo-lox66-Ptet-GFP or pR6K-Tps-lox71-Gent-lox66-Ptet-T7RP.

4.2.3 Multiple Fragments Assembly Using the *rec*ET System

- Inoculate GB05-dir single colony in fresh LB at 37 °C and incubate for 16 h.
- Dilute the overnight cultures into 1.3 mL fresh LB medium (starting OD_{600} around 0.085) and grow at 30 °C, 950 rpm for 2 h.
- Add the L-(+)-arabinose to a final concentration of 2.5 mg mL^{-1} and then grow the cells at 37 °C, 950 rpm for 40 min.
- Spin the cells at 9000 rpm for 30 s in the cooling centrifuge at 2 °C.
- Discard the supernatant as much as possible.
- Resuspend the pellet in 1 mL of ice-cold ddH_2O by a vortex.
- Spin the cells down at 10,000 rpm for 30 s in the cooling centrifuge at 2 °C.
- Discard the supernatant as much as possible.
- Resuspend the pellet in 1 mL of ice-cold ddH_2O by a vortex.
- Spin the cells down at 11,000 rpm for 30 s in the cooling centrifuge at 2 °C.
- Discard the supernatant by decanting and keep about 30 μL.

- Add 400–800 ng DNA for each fragment to the electrocompetent cells and pipette the mixture into the chilled 1 mm electroporation cuvette.
- Set the electroporator to 1350 V, 10 μF, 600 Ω.
- Gently tap the cuvette on the table to remove air bubbles and dry the metallic sides of the cuvette with a tissue.
- Place the cuvette into the holder of the electroporator, insert, and push the 'pulse' button twice.
- Add 1 mL fresh LB medium to the cuvette. Resuspend the cells gently by pipetting up and down and transfer into the reaction tube.
- Recover the cultures on a shaking incubator at 950 rpm for 60 min.
- Spread the cells on kanamycin (15 μg mL^{-1}) or gentamicin (2 μg mL^{-1}) plates, and incubate at 37 °C for 24 h.
- Inoculate single colonies in fresh LB with appropriate antibiotics at 37 °C for 16 h.
- Mini-Prep DNA and restriction analysis of the recombinants.
- The homolog arm of the target construct can be exposed by restriction digestion, and the DNA is ready for integration into chromosomal locus of the *P. luminescens* and *X. stockiae*

5 Concluding Remarks

A recombineering approach, which is mediated by phage-derived proteins either the Red operon of phage lambda or recET from the Rac prophage of *E. coli*, has many advantages over endogenous homologous recombination. Remarkably, efficient recombination with shorter homology arms circumvents the tedious work to make dedicated suicide plasmids. In line with the impact that recombineering has had on *E. coli*, Pluγβα is likely to have similar potential for modifying the chromosomes of *P. luminescens and X. stockiae*. Using such a recombineering system, the gene cluster *plu2670* of *P. luminescens* was successfully expressed by the promoter exchange approach (Yin et al. 2015b). When combined with the Cre site-specific recombination, it can be used efficiently and rapidly to generate seamless genomic modification and sequentially manipulate different genes or the locus of the genome in *P. luminescens and X. stockiae*. Knock-in or knockout mutants of multiple genes could be easily achieved by Pluγβα recombineering system. This method gives a new way to generate more genetically modified *P. luminescens, X. stockiae,* and closely related species.

A cassette with several genes can be assembled into a target construct for genome engineering of *P. luminescens* and *X. stockiae* using recET-mediated recombineering, which is an efficient in vivo homologous recombination system to join multiple DNA molecules. This strategy can be broadly applicable to any target constructs for engineering the genomes of other bacteria.

References

Balbás P, Gosset G (2001) Chromosomal editing in *Escherichia coli*. Mol Biotechnol 19(1):1–12

Bao Z, Cartinhour S, Swingle B (2012) Substrate and target sequence length influence RecTEPsy recombineering efficiency in *Pseudomonas syringae*. PLoS ONE 7(11):e50617

Bian X, Huang F, Stewart FA, Xia L, Zhang Y, Müller R (2012a) Direct cloning, genetic engineering, and heterologous expression of the syringolin biosynthetic gene cluster in *E. coli* through Red/ET recombineering. ChemBioChem 13(13):1946–1952

Bian X, Plaza A, Zhang Y, Müller R (2012b) Luminmycins A-C, cryptic natural products from *Photorhabdus luminescens* identified by heterologous expression in *Escherichia coli*. J Nat Prod 75(9):1652–1655

Bode HB, Brachmann AO, Jadhav KB, Seyfarth L, Dauth C, Fuchs SW, Kaiser M, Waterfield NR, Sack H, Heinemann SH, Arndt H-D (2015) Structure elucidation and activity of Kolossin A, the D-/L-pentadecapeptide product of a giant nonribosomal peptide synthetase. Angew Chem Int Ed 54(35):10352–10355

Brachmanna AO, Kirchnera F, Keglera C, Kinskia SC, Schmittb I, Bode HB (2012) Triggering the production of the cryptic blue pigment indigoidine from *Photorhabdus luminescens*. J Biotechnol 157(1):96–99

Bunny K, Liu J, Roth J (2002) Phenotypes of lexA mutations in *Salmonella enterica*: evidence for a lethal lexA null phenotype due to the Fels-2 prophage. J Bacteriol 184(22):6235–6249

Carter DM, Radding CM (1971) The role of exonuclease and β protein of phage λ in genetic recombination: II. Substrate specificity and the mode of action of λ exonuclease. J Biol Chem 246(8):2502–2512

Chandran S, Noskov VN, Segall-Shapiro TH, Ma L, Whiteis C, Lartigue C, Jores J, Vashee S, Chuang R-Y (2014) TREC-IN: gene knock-in genetic tool for genomes cloned in yeast. BMC Genom 15:1–11

Datsenko KA, Wanner BL (2000) One-step inactivation of chromosomal genes in *Escherichia coli* K-12 using PCR products. Proc Natl Acad Sci U S A 97(12):6640–6645

Derbise A, Lesic B, Dacheux D, Ghigo JM, Carniel E (2003) A rapid and simple method for inactivating chromosomal genes in *Yersinia*. FEMS Immunol Med Microbiol 38(2):113–116

Dong H, Tao W, Gong F, Li Y, Zhang Y (2013) A functional *rec*T gene for recombineering of clostridium. J Biotechnol 173:65–67

Frank B, Michael B (2001) LoxP-directed cloning: use of Cre recombinase as a universal restriction enzyme. Biotechniques 31(4):906–908

Fu J, Bian X, Hu S, Wang H, Huang F, Seibert PM, Plaza A, Xia L, Müller R, Stewart AF, Zhang Y (2012) Full-length RecE enhances linear-linear homologous recombination and facilitates direct cloning for bioprospecting. Nat Biotechnol 30:440–446

Fu J, Teucher M, Anastassiadis K, Skarnes W, Stewart AF (2010) Chapter eight—a recombineering pipeline to make conditional targeting constructs. In: Paul MW, Philippe MS (eds) Methods in enzymology, vol 477. Academic Press, Cambridge, pp 125–144

Glaeser A, Heermann R (2015) A novel tool for stable genomic reporter gene integration to analyze heterogeneity in *Photorhabdus luminescens* at the single-cell level. Biotechniques 59 (2):74–81

Gross F, Ring MW, Perlova O, Fu J, Schneider S, Gerth K, Kuhlmann S, Stewart AF, Zhang Y, Muller R (2006) Metabolic engineering of *Pseudomonas putida* for methylmalonyl-CoA biosynthesis to enable complex heterologous secondary metabolite formation. Chem Biol 13 (12):1253–1264

Guzman L-M, Belin D, Carson M, Beckwith J (1995) Tight regulation, modulation, and high-level expression by vectors containing the arabinose PBAD promoter. J Bacteriol 177(14): 4121–4130

Hashimoto-Gotoh T, Sekiguchi M (1977) Mutations to temperature sensitivity in R plasmid pSC101. J Bacteriol 131(2):405–412

Hu S, Fu J, Huang F, Ding X, Stewart AF, Xia L, Zhang Y (2014) Genome engineering of *Agrobacterium tumefaciens* using the lambda Red recombination system. Appl Environ Microbiol 98(5):2165–2172

Kang Y, Norris MH, Wilcox BA, Tuanyok A, Keim PS, Hoang TT (2011) Knockout and pullout recombineering for naturally transformable *Burkholderia thailandensis* and *Burkholderia pseudomallei*. Nat Protocols 6(8):1085–1104

Kimura Y, Hisano Y, Kawahara A, S Higashijima (2014) Efficient generation of knock-in transgenic zebrafish carrying reporter/driver genes by CRISPR/Cas9-mediated genome engineering. Sci Rep 4:6545

Lango L, Clarke DJ (2010) A metabolic switch is involved in lifestyle decisions in *Photorhabdus luminescens*. Mol Microbiol 77(6):1394–1405

Li D, Liu J, Guo B, Liang C, Dang L, Lu C, He X, Cheung HY-S, Xu L, Lu C, He B, Liu B, Shaikh AB, Li F, Wang L, Yang Z, Au DW-T, Peng S, Zhang Z, Zhang B-T, Pan X, Qian A, Shang P, Xiao L, Jiang B, Wong CK-C, Xu J, Bian Z, Liang Z, D Guo, Zhu H, Tan W, Lu A, Zhang G (2016) Osteoclast-derived exosomal miR-214-3p inhibits osteoblastic bone formation. Nat Commun 7:10872

Li X, Thomason LC, Sawitzke JA, Costantino N, Court DL (2013) Bacterial DNA polymerases participate in oligonucleotide recombination. Mol Microbiol 88(5):906–920

Little JW (1967) An exonuclease induced by bacteriophage λ II. Nature of the enzymatic reaction. J Biol Chem 242(4):679–686

Liu J, Berry R, Poinar G, Moldenke A (1997) Phylogeny of *Photorhabdus* and *Xenorhabdus* species and strains as determined by comparison of partial 16S rRNA gene sequences. Int J Syst Evol Microbiol 47(4):948–951

Muyrers JPP, Zhang Y, Testa G, Stewart AF (1999) Rapid modification of bacterial artificial chromosomes by ET-recombination. Nucleic Acids Res 27(6):1555–1557

Poteete AR (2013) Involvement of *Escherichia coli* DNA replication proteins in phage lambda red-mediated homologous recombination. PLoS ONE 8(6):e67440

Rainey FA, Stackebrandt E (1995) Inability of the polyphasic approach to systematics to determine the relatedness of the genera *Xenorhabdus* and *Photorhabdus*. Int J Syst Bacteriol 45(2): 379–381

Sektas M, Specht M (2005) Limited use of the Cre/loxP recombination system in efficient production of loxP-containing minicircles in vivo. Plasmid 53(2):148–163

Somvanshi VS, Sloup RE, Crawford JM, Martin AR, Heidt AJ, K Kim, Clardy J, Ciche TA (2012) A single promoter inversion switches *Photorhabdus* between pathogenic and mutualistic states. Science 337(6090):88–93

Sun Z, Deng A, Hu T, Wu J, Sun Q, Bai H, Zhang G, Wen T (2015) A high-efficiency recombineering system with PCR-based ssDNA in *Bacillus subtilis* mediated by the native phage recombinase GP35. Appl Environ Microbiol 99(12):5151–5162

Tang Z, Zhang Y, Stewart AF, Geng M, Tang X, Tu Q, Yin Y (2010) High-level expression, purification and antibacterial activity of bovine lactoferricin and lactoferrampin in *Photorhabdus luminescens*. Protein Expr Purif 73(2):132–139

Van Kessel JC, Hatfull GF (2006) Recombineering in *Mycobacterium tuberculosis*. Nat Methods 4 (2):147–152

Van Kessel JC, Hatfull GF (2008) Efficient point mutagenesis in mycobacteria using single-stranded DNA recombineering: characterization of antimycobacterial drug targets. Mol Microbiol 67(5):1094–1107

van Pijkeren J-P, Britton RA (2012) High efficiency recombineering in lactic acid bacteria. Nucleic Acids Res 40(10):e76

Wang H, Bian X, Xia L, Ding X, Müller R, Zhang Y, Fu J, Stewart AF (2014) Improved seamless mutagenesis by recombineering using ccdB for counterselection. Nucleic Acids Res 42(5):e37

Woese CR (1987) Bacterial evolution. Microbiol Mol Biol Rev 51(2):221–271

Yang P, Wang J, Qi Q (2015) Prophage recombinases-mediated genome engineering in *Lactobacillus plantarum*. Microbial Cell Fact 14(1):1–11. z

Yin J, Hoffmann M, Bian X, Tu Q, Yan F, Xia L, Ding X, Francis Stewart A, Müller R, Fu J, Zhang Y (2015a) Direct cloning and heterologous expression of the salinomycin biosynthetic gene cluster from *Streptomyces albus* DSM41398 in *Streptomyces coelicolor* A3(2). Sci Rep 5:15081

Yin J, Zhu H, Xia L, Ding X, Hoffmann T, Hoffmann M, Bian X, Müller R, Fu J, Stewart AF, Zhang Y (2015b) A new recombineering system for *Photorhabdus* and *Xenorhabdus*. Nucleic Acids Res 43(6):e36

Zhang Y, Buchholz F, Muyrers JPP, Stewart AF (1998) A new logic for DNA engineering using recombination in *Escherichia coli*. Nat Genet 20(2):123–128

Zhang Y, Muyrers JPP, Testa G, Stewart AF (2000) DNA cloning by homologous recombination in *Escherichia coli*. Nat Biotechnol 18(12):1314–1317

MIX
Papier aus verantwortungsvollen Quellen
Paper from responsible sources
FSC® C105338

If you have any concerns about our products,
you can contact us on
ProductSafety@springernature.com

In case Publisher is established outside the EU,
the EU authorized representative is:
Springer Nature Customer Service Center GmbH
Europaplatz 3, 69115 Heidelberg, Germany

Printed by Libri Plureos GmbH
in Hamburg, Germany